Extreme Events
A Physical Reconstruction and Risk Assessment

The assessment of risks posed by natural hazards such as floods, droughts, earthquakes, tsunamis or tropical cyclones is often based on short-term historical records that may not reflect the full range or magnitude of events possible. As human populations grow, especially in hazard-prone areas, methods for accurately assessing natural hazard risk are becoming increasingly important.

In *Extreme Events* Jonathan Nott describes the many methods used to reconstruct such hazards from natural long-term records. He demonstrates how long-term (multi-century to millennial) records of natural hazards are essential in gaining a realistic understanding of the variability of natural hazards likely to occur at a particular location. He also demonstrates how short-term historical records often do not record this variability and can therefore misrepresent the likely risks associated with natural hazards.

This book will provide a useful resource for students taking courses covering natural hazards and risk assessment. It will also be valuable for urban planners, policy makers and non-specialists as a guide to understanding and reconstructing long-term records of natural hazards.

JONATHAN NOTT is Professor of Geomorphology at James Cook University in Queensland, Australia. His broad research interests are in Quaternary climate change and the reconstruction of prehistoric natural hazards. Other research interests include long-term landform evolution, plunge pool deposits (terrestrial floods) and reconstructing tropical cyclone climatology from deposits of coral shingle and shell. He is a member of the National Committee for Quaternary Research, Australian Academy of Science. His research has been published in many international journals including *Nature; Earth and Planetary Science Letters; Geophysical Research Letters; Journal of Geophysical Research; Marine Geology; Palaeogeography, Palaeoclimatology, Palaeoecology; Geology; Journal of Geology; Quaternary International; Journal of Quaternary Science; Quaternary Science Reviews; Environment International*; and *Catena*.

Extreme Events

A Physical Reconstruction and Risk Assessment

JONATHAN NOTT

CAMBRIDGE UNIVERSITY PRESS
Cambridge, New York, Melbourne, Madrid, Cape Town, Singapore, São Paulo

Cambridge University Press
The Edinburgh Building, Cambridge CB2 2RU, UK

Published in the United States of America by Cambridge University Press, New York

www.cambridge.org
Information on this title: www.cambridge.org/9780521824125

First published 2006

Printed in the United Kingdom at the University Press, Cambridge

A catalogue record for this publication is available from the British Library

ISBN-13 978-0-521-82412-5 hardback
ISBN-10 0-521-82412-5 hardback

To Monkey, Blue Eyes and Curly Tops

Contents

Acknowledgments

I would like to thank Drs Scott Smithers and James Goff for their constructive comments on this manuscript. Their assistance was invaluable. The views expressed in this book are entirely mine.

I would also like to thank the following for permission to reproduce in part and/or full the following:

Elsevier Science for Figures 2.1, 2.8, 2.9, 5.9, 6.4, 6.5, 7.4, 7.5, 7.6, 7.7, 7.8, 8.2, 8.4.

Nature Publishing Group for Figures 2.4, 2.5, 2.6, 2.10, 4.3, 4.11.

Geological Society of America for Figures 2.2, 7.2, 7.3.

American Geophysical Union for Figures 5.7, 7.9, 7.10, 8.3.

Coastal Research Foundation or Figure 2.4.

National Atmospheric and Oceanic Administration for Figure 2.11, 2.12.

Professor D.A. Kring for Figures 9.1, 9.2, 9.3, 9.4, 9.5.

Professor S. Bondevik for Figure 5.8.

Professor V. Baker for Figure 3.2.

1

Introduction

The problem with natural hazard risk assessments

There is a problem with many natural hazard risk assessments. They do not incorporate long-term and/or prehistoric records of extreme events; otherwise known as natural hazards when they affect humans physically, psychologically, socially or economically. Short historical records are frequently assumed to be a true reflection of the long-term behaviour of a hazard. Historical records may be appropriate, in this regard, where they extend for at least several centuries or even a millennium such as in China. However, in many countries, like the United States, United Kingdom and Australia, the historical record is often not much longer than 100 years. Many assessors of risks from natural hazards see these short records as appropriate for determining the natural variability of a hazard. From this they extrapolate to determine the magnitude of less frequent, higher magnitude events and construct probability distributions of the occurrence of a hazard at various return intervals. Inherent in this process is the assumption that natural hazards occur randomly over a variety of time scales and that the mean and variance of the hazard do not change. This may be true in certain circumstances, especially shorter time periods, but is often not the case for longer intervals. When we rely upon short historical records we run the real risk of not capturing the natural variability of the hazard. Here lies the crux of the problem – when we do not understand the true nature of the hazard in question we cannot hope to make realistic assessments of community vulnerability and exposure and we increase the chance of making an unreliable estimate of the risk of that hazard. Our ability to increase community safety and reduce economic loss is dependent upon our understanding of the behaviour of the hazard. Short historical records rarely display sufficient information for us to interpret this natural variability. Nature, however, effectively records its own extreme events, and often its not so extreme ones, providing us with a documented history in the form of natural records that can display the full range of variability of most hazards that confront society. Long-term records, therefore, are the only real source for uncovering the true nature of the behaviour of natural hazards over time.

Scientists who study the Quaternary – the most recent period of geological time (or approximately the last 2 million years) know that natural events including hazards often occur in clusters or at regular to quasi-periodicities. Over longer time intervals, the Quaternary record shows us that the periodicity of events, including climatic changes, is governed by many factors, some of which are external to the Earth. For example, climatic changes of various scales occur at intervals from 100 000 years to 11 years based upon regular variations in the orbit of the Earth around the Sun, along with the tilt of the Earth's axis and precession of the equinoxes, to sunspot activity. There are many other regular cycles of climate change that occur in between the 100 000 and 11 year cycles that have been uncovered from a variety of natural records. The clear message from Quaternary records is that many of the climatic changes that occur on Earth do not occur randomly and in this sense are not independent of time. The same could be expected of many extreme natural events.

While it is true that natural records do not document the event as accurately as the instrumented record, they are nonetheless of sufficient precision to show us the magnitude of the most extreme events and how often these events are likely to occur. Even more importantly, natural records provide us with a very effective means by which to test the assumptions of the stationarity of the mean and variance, and randomness of occurrence of a natural hazard. In the absence of these tests we cannot hope to realistically assess community vulnerability and exposure, and therefore risk. Social scientists, involved in that part of the risk evaluation process devoted to community and social parameters, also need to be aware of the assumptions made by scientists and engineers in determining the physical nature of the hazard. This is frequently not the case and planners are left with false impressions of which areas are safe for urban, industrial and tourism developments.

We can only really attempt to reduce risk from natural hazards when we factor the dependence on time of a hazard into our risk equations, or at least test for it, and then see where the historical record fits in the sequence. Unfortunately, this approach is rarely adopted. It is imperative that we examine each of the forms of evidence, being the instrumented, historical and prehistoric records when undertaking risk assessments. Many practitioners, however, are unfamiliar with prehistoric records and are hesitant to incorporate them into risk assessments. Through familiarity, though, comes awareness of the insights that the prehistoric record can provide into gaining a more realistic impression of the behaviour of natural hazards.

The risk assessment process

Risk from natural hazards is a function of the nature of the natural hazard (i.e. probability of its occurrence), community vulnerability and the elements at risk. Risk can have a variety of meanings and is sometimes used in the sense of the probability or chance that an event will happen within a specific period of time. Alternatively, risk can refer to the outcomes of an event occurring. In this latter sense, risk refers to the expected number of lives lost, persons injured, damage to property and disruption of economic activity due to a particular natural phenomenon. Risk is really the product of the specific risk and the elements at risk. The specific risk here means the expected degree of loss due to a particular natural phenomenon and is a function of both the natural hazard and vulnerability (Fournier d'Albe, 1986). The elements at risk, otherwise known as the level of exposure, refers to the population, buildings, economic activities, public services, utilities and infrastructure that may be directly impacted by the hazard.

Community vulnerability is determined by the social and demographic attributes that influence a person's perception of the risk to the hazard. It often concerns peoples' attitudes, preparedness and willingness to respond to warnings of an impending hazard. Anderson-Berry (2003) notes that community vulnerability is not a static state but a dynamic process. It is generated by the complex relationships and inter-relationships arising from the unique actions and interactions of the social and community attributes and characteristics of a particular population.

These attributes and characteristics include:

- societal structures, infrastructure and institutions including the integrity of physical structures;
- community processes and structures such as community organisation, mobility of the household population, and community cohesiveness and the social support this affords; and
- demographic and other characteristics of individuals within the community such as age, ethnicity, education and wealth (Keys, 1991; Fothergill, 1996; Buckle, 1999; Fothergill *et al.*, 1999; Cannon, 2000).

These factors, along with actual experiences of the hazard in question, help to shape the individual's and the community's perception of risk. Anderson-Berry

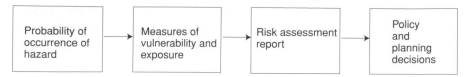

Figure 1.1. Generalised sequence of process occurring in a risk assessment. In reality there are many feedback loops between the steps outlined here.

(2003) notes that people individually and collectively decide what precautionary measures will be undertaken and how warnings will be complied with so as to ensure that the loss resulting from a hazard event is limited to an acceptable level. If the perceived risk is a true reflection of the actual risk associated with a particular hazard, then mitigation strategies, warning compliance and response preparedness are likely to be appropriate and vulnerability can be minimised. If risk perception is biased, the reverse is true and vulnerability may be increased. The perception of risk, therefore, can often be the precursor to determining the level of exposure to that risk, although it is true that perceptions can and do change over time. So increasing awareness with time, due to education about or experience with the hazard may result in the realisation that more elements are exposed than previously thought. The level of exposure, therefore, is a function of past and present perceptions of risk.

The total risk is often expressed as:

$$\text{risk (total)} = \text{hazard} \times \text{elements at risk} \times \text{vulnerability}$$

Often each of these components of the risk equation is determined separately and in isolation from the other components. When this occurs it is typically a function of the background and the training of those employed to undertake the task. For example, it is common to have a physical scientist or engineer determine the probability of occurrence of the natural hazard, whereas social scientists are usually best trained to deal with vulnerability and exposure. The two sometimes do not fully comprehend each other's assessments and will not question the veracity of the methods used or the results obtained. The social scientist, for example, may not feel comfortable reviewing the engineer's assessment of the hazard and will accept, at face value, the results as being correct or the best that can be obtained. The level of exposure is then assessed which in turn influences the assessment of vulnerability and vice versa. A report is often produced which becomes the basis for planning and policy decisions by various levels of government. Hence, the engineer's assessment is critical to and underpins all subsequent stages of the risk assessment process. Figure 1.1 outlines this process.

Each of these stages is critical and no less valuable than the others in terms of reducing risk from natural hazards. However, any variation to the outcome of the first stage (i.e. the hazard probability) influences each of the other stages; hence, each of the latter are dependent on the former. For example, hundreds to thousands more homes may be deemed to be exposed to tsunami inundation depending upon whether the assessed probability of occurrence of tsunami run-up height is 1 or 2 m above a certain datum for a given time interval along a densely populated low-lying coast. Likewise, government policy decisions may set aside a considerably larger area of coastal land deemed to be unsuitable for permanent development depending upon the height of the tsunami run-up determined. Obtaining the most accurate and realistic estimate of the magnitude of a hazard at a given probability level, therefore, is usually in the best interests of all concerned in the risk assessment process, and likely even more so for those potentially subject to impact by the hazard. The perception of the most realistic estimate, however, can vary and is the essence of the earlier stated problem in the risk assessment process. There can be a difference between the mathematical and/or statistical certainty of a certain magnitude hazard occurring in a given time period and the so-called realistic estimate of the size of that hazard.

Mathematical and statistical certainties versus realistic estimates

The probability of occurrence of a given event is a statistical measure. Probability assessments are normally based upon the assumption that the event occurs randomly with respect to time, and that events occur randomly with respect to each other. Randomness in this sense is commonly likened to the probability of obtaining a head or tail in tossing a coin. We determine that there is always a 50% probability of obtaining a head or tail each time we toss the coin. Each toss occurs independently of the other and hence the outcome of the toss is random with respect to past tosses and therefore time. This does not mean of course that we will get a head, then tail, then head with each successive toss. Time is a dependent factor when we consider that with increasing time or number of tosses we increase the probability of obtaining two or more heads in a row. But if we take any specified period of time we can expect to get a certain outcome based upon the independence of tosses relative to each other and the outcome is a function of randomness. The same view is taken with respect to the occurrence of many, but not necessarily all, natural hazards. Each year in the time series, in a sense, represents a toss of the coin. For a given magnitude event we could expect a certain probability of occurrence of that hazard. With increasing periods of time this event will have a greater probability of occurrence (like obtaining two or more heads in a row) but its probability of occurrence in

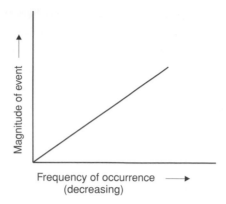

Figure 1.2. Typical relationship between event magnitude and frequency. Note that larger magnitude events occur less frequently than smaller magnitude events.

any one year always remains the same. In fact, its probability of occurrence is regarded as remaining the same for any period of time. This remains the case if the event occurs independently of prior events and therefore, like the toss of a coin, occurs randomly within any given period of time. If external factors influence the occurrence of that event then it will not be occurring independently of a specified period of time. To once again use the coin tossing analogy, if something influences the outcome of the tosses of a coin for some specific period of time then we can no longer say that the outcome is a random occurrence. We know that this is unlikely to be the case when tossing a coin, but it is a very real possibility with respect to the occurrence of natural hazards over time.

There are statistical measures, such as Bayesian analyses, which do assume that prior events influence the occurrence of subsequent events and these are sometimes used in risk assessments of some natural hazards such as earthquakes. Bayesian analysis is usually only used for hazards that have known build-up and relaxation times, as occurs for example when crustal stresses along a fault or tectonic plate boundary build to the point of release causing an earthquake. Following the earthquake, it takes some time for those stresses to once again build to a level to induce the next earthquake. So the time between earthquakes is not random and the probability of an earthquake occurring at that specific location is reduced for a period and hence varies over time. This can be called a conditional probability. Other hazards, such as many atmospheric hazards, however, where such processes are thought not to operate, are generally regarded as occurring randomly with respect to previous events and in this sense randomly with respect to time.

High-magnitude events usually occur less frequently than low-magnitude events. A typical distribution of events over time is shown in Figure 1.2. It is

Table 1.1 *Percentage probabilities of occurrence for given time intervals*

Return period (years)	Annual exceedence probability (AEP) (%)	Probability of occurrence (%) for the period (years)				
		25 years	50 years	100 years	200 years	500 years
50	2	39	63	87	98	99.9
100	1	22	39	63	87	99.3
200	0.5	12	22	39	63	92
500	0.2	5	9.5	18	33	63
1000	0.1	2.5	5	9.5	18	39
5000	0.02	0.5	1	2	4	9.5

more likely, therefore, that a place will experience a high-magnitude event with increasing time. So location X, for example, is unlikely to experience a high-magnitude event over 100 years but is reasonably likely to experience this event over a 1000 year period. But, as stated earlier, the likelihood of that high-magnitude event occurring in any 100 year period remains the same, even if it has been 900 years since the last high-magnitude event. The probability of that event occurring between year 900 and year 1000 is exactly the same as the probability of occurrence between year 100 and year 200. Probabilities, therefore, are determined according to the time interval to which they pertain. The probability of the 1 in 100 year event (1% annual exceedence probability, AEP) occurring is 1/100 in any given year. In other words, this event has a 1% chance of occurring in any given year. Likewise, it has a 39% chance of occurring in a 50 year period, 63% chance of occurring in a 100 year period and 99.3% chance of occurring in a 500 year period. Table 1.1 sets out the probabilities of events occurring over various time intervals. The determination of these probabilities is calculated according to the binomial distribution. The equation for the binomial distribution is

$$P(r) = {}^{n}C_{r}p^{r}q^{n-r} \tag{1.1}$$

where ${}^{n}C_{r}$ (the binomial coefficient) $= n!/r!(n-r)!$ and where $P(r)$ is the probability of occurrence, n is the number of events in the record, $r = 0$ and $q = 1-p$.

The binomial distribution is based upon the randomness of occurrence of events over time. The same distribution is used to explain the chance of obtaining a head or tail in the toss of a coin which is most certainly a random event. By applying the same statistical probability distribution to the occurrence of natural hazards we make the assumption that these events occur randomly like the chance of obtaining a head or tail in the toss of a coin.

The application of this approach to determining the probability of occurrence of a natural hazard is shown in Figure 1.3. The majority of events in Figure 1.3

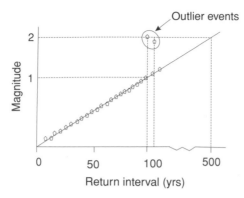

Figure 1.3. Magnitude–frequency curve and outlier points.

can be seen to fall roughly on a straight line; however, two events do not. These two events are referred to as 'outliers' and they are of higher magnitude than any other events to have occurred over the past 100 years. The probability distribution suggests that despite the fact that these outlier events have occurred within the last 110 years they do not belong to the normal range of events that could be expected to occur within this time frame. By extending the line representing the magnitude–frequency relationship for this particular hazard, these events appear more likely to correspond to the approximately 1 in 500 year event (0.2% AEP). Such a conclusion is firmly based upon the assumption that the slope of the line representing the magnitude–frequency relationship is applicable to any 100 year period. Therefore, this line, which only covers events from the last 100 years, is typical of any 100 year period. When we make this assumption we also deem it safe to extrapolate this line to determine an accurate estimate of the size of less frequent events. Whether this is a realistic interpretation of the nature of the natural hazard, however, is rarely ever tested during the risk assessment process. Nor is the possibility that non-stationarity may be evident when longer time series or records of events are examined. In the above situation stationarity has been assumed. Unfortunately though, nature rarely displays stationarity over the long term.

Stationarity in time series

Stationarity occurs when the relationship between the magnitude and the frequency of an event and/or its variance remains unchanged with time. Non-stationarity refers to a condition where the relationship between the magnitude and frequency and/or variance changes over time. In the former case this can be reflected as a change in the slope of the magnitude–frequency line. In these situations the magnitude of a certain frequency hazard changes. If the slope of

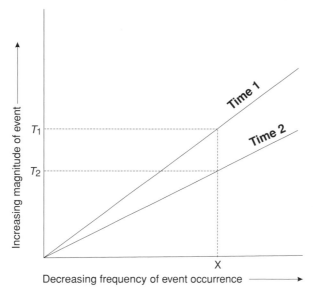

Figure 1.4. Non-homogeneity of magnitude–frequency relationship over time.

the line steepens, events of greater magnitude will occur more frequently. The converse is true when the slope of the line decreases. These changing relationships are shown in Figure 1.4. Two lines are presented here. Each line represents a different relationship between the frequency of occurrence of an event and its associated magnitude. An event with frequency X will have a magnitude of T_1 during a period named here as Time 1, and magnitude T_2 during the period Time 2. These periods may be years, decades, centuries or millennia in length. This is non-stationarity (see Fig. 1.5).

Stationarity, therefore, can only be assumed to occur for limited periods of time. The length of these periods is variable, and they will often dominate an entire short historical observational record. Hence, these records will not display non-stationarity even though this is the normal behaviour of a hazard over the longer term. Planners might say that the length of the period in question (i.e. the current period of stationarity) is longer than the proposed planning cycle so what is the use of attempting to test for non-stationarity. However, if we choose not to recognise that these changes occur we will not seek explanations for the cause of these changes and without knowing the causes we will not endeavour to understand the changes. Hence, we will not know when a change is likely to occur. If such changes do occur within the planning period, and because stationarity is assumed, the occurrence of a high-magnitude event will be regarded as an outlier, or an event of much lower frequency. This, of course, has implications for insurance premiums and claims, and future policy and planning. It also affects the way we perceive risk and, therefore, influences

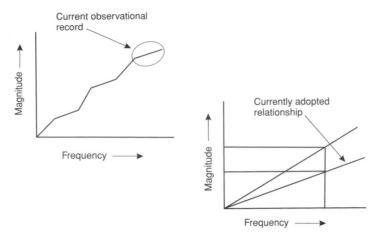

Figure 1.5. Non-stationarity in the time series.

community vulnerability and levels of exposure. If we believe that the high-magnitude event is an outlier then we will believe that the chances of an event of similar size occurring during the next planning cycle, or the near future, are low (i.e. as set out in Table 1.1). For example, if we perceive the outlier event to be a 1 in 500 year event (0.2% AEP) then we will draw the conclusion that its chances of occurring in the next 50 years are less than 10%. However, if we find from a longer-term record that this magnitude event is really more likely to be a 1 in 100 year event (and indeed it did occur only a little over 100 years ago in reality) then its chance of occurring in the next 50 years is about 39%. Alternatively, it could be an event that signals the onset of a change in hazard behaviour. Either way, when we assume it was a 1 in 500 year event, and ignore the possibility that it is part of a normal series of events for a regime that was not recognised due to the brevity of the observational record, then we will have placed potentially large numbers of people and property at higher than pre-dicted levels of risk. Each change of phase in hazard behaviour, i.e. change of slope in the magnitude–frequency curve, can be referred to as a hazard regime (Nott, 2003). These regimes can be likened to alternating periods of variable sta-tionarity. Recognising the possibility that these regimes may exist for a hazard at any location can only help us to gain a more realistic view of the nature of the hazard.

Reality versus reasonableness

Mathematics and statistics have dominated our approach to assessing risk from natural hazards. The strength of these methods is in their predictive capabilities. But as Baker (1994) states, prediction can be powerfully wrong as well as powerfully right. For many natural hazard risk assessments, determination of the nature of the hazard, being the first part of the risk assessment process, is dominated by an engineering and physics approach. This is especially the case with floods, droughts and tropical cyclones. The reasoning process of physics is analytical, meaning that first principles are assumed and consequences deduced according to structured logic, which is often mathematical. This deduction is often thought of as a prediction, but these are not prophecies of future events. Predictions are logical deductions whose accuracy, or contact with reality, is in their match with a measured property of nature (Baker, 1994). The match can be in the form of an experiment, which in some situations can be based upon the results of numerical models. Many models, however, are based upon first principle assumptions concerning the nature of the hazard. Testing these models cannot realistically be undertaken by comparisons with observations of future events because the suggested answers to the problem have been used as the basis of constructing the model, which is in turn being used to derive a solution to the same problem. This, therefore, is a circular argument.

Assessing the probability of occurrence of tropical cyclones is a case in point. The return period of extreme cyclone-generated storm tides is often calculated by simulating a large number of storm tides from cyclones randomly selected from frequency distributions of observed (historical) cyclone characteristics. Randomly selected cyclones are then run through a numerical storm tide model. The total number of modelled events represents a number of years that depends on the observed cyclone frequency for a region. For example, if cyclones occur on average once every 2 years at a particular location each storm tide simulation represents 2 years' data (1000 simulations = 2000 years' data). The storm tide simulations are ranked in decreasing order (highest = 2000 year event, 2nd = 1000 year event, 3rd = 667 year event). In Australia, the cyclone frequency analysis is confined to the last 40 years (since 1960) because detailed instrumented data only exist for this period. In the Cairns region of northeast Queensland, for example, cyclones have occurred on average once every 10 years over the last 40 years. Based on this recurrence interval, 1000 storm tide simulations can be deemed to be equivalent, at the determined frequency of 1 cyclone per 10 years, to 10 000 years of storm tide simulations. In this instance the 100 year storm tide = 2.15 m above Australian height datum (AHD) + wave set-up (~0.5 m) and the 1000 year storm tide = 3.75 m (AHD) + wave set-up. Based on this analysis Cairns can expect a large storm tide from a Category 5 cyclone (~900 hPa) approximately every 1000 years.

The approach adopted here is mathematically and statistically rigorous and based upon sound physics. However, an assumption is made that the 10 000 years of storm tide simulations actually reflects what would most likely occur, in reality, over a 10 000 year period. When this assumption is based upon only decades of observational and/or instrumented data, or even a century of data, then it is clear that it is unlikely to be realistic because it assumes stationarity of the record. And therefore, the accuracy of this hypothesis is not tested. Problems arise when planners and policy makers accept the results as realistic or at least the most realistic achievable. The only realistic test of this assumption is to make comparisons against events that have actually occurred during the time period supposedly covered by the model runs i.e. 10 000 years. Many modellers might say that this is not achievable because we do not have 10 000 years of continuous records of tropical cyclone storm tides. We do, however, have geological records of the most extreme events over the last 5000 years. It is true that these records do not present data with the same level of precision as an instrumented record. However, such records can be used to test the accuracy of model simulations to within at least an order of magnitude. This has to be a considerably better option than leaving such assumptions about the levels of risk from the hazard untested. The geological record, for example, can tell us the height of the 1 in 3000 year storm tide event based upon the elevation of deposits of marine debris such as corals and shell above normal sea-level and this can be compared to the simulated 3000 year storm tide event. Hence, we do not need to rely solely upon deduction i.e. the use of simulated events can complement this approach through retroduction or the examination of a record of events that really occurred over this length of time. In this sense, the prehistoric record is more than sufficiently accurate to perform this task.

Prehistoric records focus on the contingency of past phenomena. Contingency holds that individual events matter in the sequence of phenomena (Baker, 1994). Changing one event in the past causes the sequence of historical events to be different from that which is observed to have occurred. One recognised extreme event only decades before an historical record begins changes the way we would view that historical record. The fact that this event occurred in time and space is real and it is registered in the prehistoric record. It is only our assumption about nature's behaviour that holds that such an event might be an outlier because these assumptions are formulated in the absence of any knowledge of these prehistoric events. If they were to have occurred within the observed, and particularly instrumented, record they would be more readily accepted as a real event and form part of the assumption of the statistical distribution of events. Baker notes that such uniqueness to history is alien to the thinking of physicists who seek timeless, invariant laws presumed to be fundamental for nature. This same view is held by engineers who often, in terms of hazard risk assessment, develop conceptual idealisations about nature. Physics is the science devoted to discovering, developing and refining those aspects of reality that are amenable to mathematical analyses (Ziman, 1978). It is powerful science and has been at the forefront of human intellectual achievement for many decades, but it does not traditionally take a historical view of nature. The geological approach to examining nature on the other hand is one of *'taking the world as it is'* which can be thought of as a type of realism (Baker, 1994). The focus in geology is in deriving hypotheses from nature rather than applying elegant theories. Baker suggests that the geological approach involves synthetic reasoning according to classical doctrines of commonsensism, fallibilism and realism. The geology approach does not seek to provide answers to puzzles about the sequence of past and possibly future events by imposing limiting assumptions upon the real world. Too often engineers impose limitations upon their investigation of the nature of natural hazards because they do not feel comfortable adopting a geological and/or historical approach to the problem. By using prehistoric data as a test of the initial assumptions made in natural hazard risk assessments, we adopt a more scientific approach to the problem. In other words, we seek to discover nature through research and not halt our attempts at discovery by adopting limiting assumptions. Prehistoric data is a powerful ally in the quest to reduce risk from natural hazards as it helps to test these initial assumptions. Technically though, the prehistoric data does not always record natural hazards because these events only become a hazard when they affect humans. If they occur in isolation from humans they can be better referred to as extreme natural events. Extreme events, as registered in prehistoric records therefore, are real as they have occurred in a temporal and spatial sequence that is discoverable.

Concluding comments

The dilemma presented here exists because different professionals, usually engineers and natural scientists (often geo-scientists), see the nature of hazards from different perspectives. The engineering approach is to develop a model of this behaviour which is seen to remain constant over time. The natural scientist's view is that the behaviour of hazards may change over time and no one particular slice of time is necessarily reflective of the nature of that hazard. Engineers typically regard the evidence from natural long-term records as too imprecise to be meaningful. So they are often not interested in incorporating such information into their models. However, the natural scientist's view is that while this evidence may be less precise than the engineer's data, it is nonetheless more than sufficiently accurate to test the assumptions upon which the engineer's model rests. This scenario and its typical consequences can be explained by a simple story.

Imagine a community of intelligent bugs that live on the sides of a smooth sided dish shaped pond. They are a complex community with many specialist occupations. They have been living in the pond for about 100 years. The principal natural hazard facing their society is variations in water levels in the pond. The engineer bugs monitor the water levels with precision instruments and have done so for nearly 100 years on a daily basis. They have a lot of data. The scientist bugs are interested in the deposition of sediments around the sides of the pond and they recognise that the position of the sediments relates to the level of water in the pond at the time of deposition. The scientist bugs are also interested in the chemical composition of the sediments and they use this information to monitor pollution levels emitted by the bug community. The sediments around the sides of the pond can be likened to a dirt ring around a bath tub. Few would question that the height of the dirt ring is a reflection of the water level at the time of sediment deposition, but the dirt ring does not record those water levels as accurately as an instrument.

The scientist bugs observe another sequence of sediments at a much higher elevation on the sides of the pond. They conclude that these sediments were deposited by considerably higher water levels. The engineer bugs agree and tell the scientist bugs that their model shows that this must have been the 1 in 500 year (0.2% AEP) event. The scientist bugs collect samples from the higher level sediments for radiocarbon analyses which return ages of around 150 years. The engineer bugs say that the sediments still represent the 1 in 500 year event and that it is an outlier. The scientist bugs wish to test this hypothesis so they examine a range of higher level sediments around the sides of numerous ponds throughout the region. In the end they manage to date events spanning the last 1000 years. The average return interval for the higher level events is 90 years. They present their data to the engineer bugs who dismiss it as an inaccurate measure of past water levels. They still say that the higher water levels represent the 500 year or even less frequent events.

A dilemma exists. Both groups see the behaviour of this hazard from different perspectives. The engineer bugs regard accuracy of measurements as a guide to the veracity of their hypothesis. The scientist bugs wish to test this hypothesis or model outcome and suggest that it is not necessary to have data as precise as that derived from instruments to do this. They say that despite the fact that they can only determine the water level to within 10–20% accuracy, the presence of the higher level sediments together with the radiocarbon ages with their uncertainty margin of a decade, demonstrate that the initial hypothesis is likely flawed and needs to be reviewed. They can show that these events around the region definitely do not have a return interval of 500 years. They argue that the sediments represent real events that actually occurred and the model only predicts events of that magnitude to occur within the 500 year return interval. The scientist bugs have adequately tested the hypothesis. But this does not dismiss the engineers' data or approach in terms of development of a model. The two groups can work together by incorporating the long-term records into the numerical models of the relationship between water levels and time. Unfortunately, in real life this often does not happen, especially in many countries dealing principally with atmospheric hazards. With geological hazards though, such as earthquakes, substantial progress is being made in this regard.

Aims and scope of this book

The problem with natural hazard risk assessments is their lack of incorporation of prehistoric data on extreme events. Part of this problem no doubt stems from a lack of familiarity with this form of record by those often left to undertake the risk assessments. This information, while extensive and detailed in terms of both results from studies and the methodologies employed to derive these records, is presently largely housed and scattered within a variety of technical journals. The aim of this book has been to synthesise this information into a single volume so that practitioners and students of natural hazard risk assessments can understand the methods used to derive, and the value of using, prehistoric records of extreme events. The overriding message here is that we cannot afford to omit natural records of extreme events from risk assessments of natural hazards. To do so potentially places people's lives, homes and investments at much greater risk than otherwise needs to be the case. This book is a first step towards changing a culture that tends to exclude such information because of a certain level of naivety about its value.

The chapters are arranged according to the type of extreme event. The first part of each chapter deals with the mechanisms causing extreme events such as the processes causing earthquakes. The impacts of these events historically upon humans are also detailed for each type of hazard. In this sense, the start of each chapter deals with topics covered by several excellent books on natural hazards (Alexander, 1993; Bryant, 2005). But the latter part, and indeed the majority of each chapter, is unlike standard texts on hazards as it deals with the prehistoric records of these extreme events. The numerous methods used to derive these records are outlined, and examples and results of a number of published studies examining these records are reviewed. The aim has not been to provide an exhaustive review of the many studies of prehistoric extreme events, but rather to outline the methods by which these records are derived and in so doing provide examples of some of the studies of these extreme events. The final chapter examines the nature of extreme events over time and demonstrates that these events do tend to cluster over various periods and do not, therefore, occur entirely randomly. This is the crux of the limiting assumptions made in so many natural hazard risk assessments. Hopefully, the synthesis of this book as presented in the final chapter will convince many practitioners of the need to adopt a geological and long-term historical approach, together with the mathematically rigorous approaches of engineering and physics, to reducing risk from natural hazards.

2

Droughts

Drought may be defined as an extended period of rainfall deficit within an otherwise higher rainfall regime. However, this definition is subjective, for the length of period without rain varies between locations depending on major climatic regimes. Periods of rainfall deficit deemed to be a drought are also, in many cases, based upon rainfall patterns observed over relatively short historical records. In a country with modern records extending back only one to two centuries, a period of five years with a rainfall deficit may be termed a drought despite the fact that such events may have extended naturally to 50 years or more prior to the start of the modern record. The severity of a drought, therefore, is largely determined by its impact on humans rather than an arbitrary measure such as the Richter scale with earthquakes. Droughts affect both agricultural and urban communities but it is often the former that feel the effects before and more severely than the latter.

Drought is one of four classes of water scarcity (Robinson, 1993). These are:

(1) aridity, a permanent shortage of water caused by a dry climate;
(2) drought, an irregular phenomenon occurring in exceptionally dry years;
(3) desiccation, a drying of the landscape, particularly the soil, resulting from activities such as deforestation and over-grazing; and
(4) water stress, due to increasing numbers of people relying on fixed levels of run-off.

Drought may also be seen to be a temporary form of water scarcity, aridity a permanent form and desertification a human-induced form (that may be either temporary or permanent). The characteristics and causes of these forms of water scarcity are listed in Table 2.1 after Chapman (1999).

Droughts are often caused by short-term climatic changes relating to latitudinal shifts in the position of the intertropical convergence zone (ITCZ), the Hadley cells, the jet stream and alternations in the El Niño Southern Oscillation (ENSO). The ITCZ is the region usually occurring between the Tropics of Cancer and Capricorn, depending on the season, where tropical air rises due to intense heating from the Sun. The rising air spreads northward and southward towards the poles where it cools and descends at 20°–30° north and south of the equator. Where the sinking air moves back towards the equator the circulation pattern is known as the Hadley cell. Some air also moves towards the poles and is uplifted by frontal systems associated with the subpolar low pressure systems. This circulation is known as the Ferral cell. The jet stream is an upper troposphere wind which often forms a looping pattern as it encircles the globe. In the northern hemisphere, the jet stream forms a wave-like pattern of alternating high and low pressures in a line between Siberia across the north Pacific to central North America. This looping pattern is known as a Rosby wave and is quasi-stationary and orographically fixed in position by the Tibetan Plateau and the Rocky Mountains (Bryant, 2005). The amplitude of the Rosby wave is defined by the pressure differences between the respective high and low pressure cells and its amplitude increases with increasing pressure in the pressure cells. Rosby waves can affect the latitudinal position of the jet stream, which in turn can cause the passage of high pressure systems and reduced precipitation into areas that normally receive higher annual rainfall. Such changes to the looping pattern of the jet stream can result in both short-term drought and semi-permanent climatic change (Bryant, 2005)

Drought conditions are also strongly influenced by the El Niño Southern Oscillation. El Niño events occur when sea surface temperatures along the coast of South America increase over months to several years. Normally, warm surface waters dominate the Western Pacific and cool surface waters occur in the Eastern Pacific. This pattern is maintained by the generally easterly trending trade winds (north-easterlies in the North and south-easterlies in the South Pacific Oceans). These trade winds are related to the annual migration of the Hadley cells following the seasonal movement of the sun southward and northward of the equator (ecliptic). High pressure systems normally develop over the equatorial ocean west of South America and over the Peruvian coast. Easterly trade winds occur when unstable air rises over the Indonesian–Australian region and cooler air descends over the central to eastern Pacific region. Surface waters are blown back across the Pacific where they pile up in the Coral Sea off northeastern Australia. During an ENSO event this system breaks down with a weakening of the easterly trade

Table 2.1 *Aridity, drought and desertification (from Chapman, 1999)*

Drought (temporary)	Aridity (permanent)	Desertification (human-induced)
Can occur in any climate	Climatic state associated with global circulation patterns	Resulting from:
Unpredictable occurrence	Low annual precipitation normal	Mining of groundwater
Uncertain frequency	Rainfall highly variable in both time and space	Over-grazing and unwise cultivation
Uncertain duration	High evaporation normal	Attempts to extract more from land than natural productivity allows
Uncertain severity	High solar energy input	Unwise irrigation practises leading to salinisation, Producing symptoms of:
Long periods of lower than average precipitation	Large annual temperature variations normal	Reduction of perennial vegetation cover
Protracted periods of diminished water resources	Low productivity of natural ecosystem normal	Aquifer depletion, land subsidence
Diminished productivity of natural ecosystems	Low productivity of farms and rangeland normal	Damaged surface soil and subsoil, decreased infiltration
Diminished productivity of farms and rangelands	Sparse human settlement normal	Increase in soil temperature
Deterioration of farm land and range land		Compaction and salinisation of soils
		Oxidation of soil organic matter
		Reduction of water-holding capacity of soil
		Increased propensity for flash flooding and further erosion
		Loss of productivity of natural ecosystems and of farms and rangeland
		Raised surface albedo (tends to diminish rainfall)
		Invasion of former farms and rangeland by woody weeds

winds. Convection of air increases in the central to eastern Pacific Ocean followed by a rise in sea surface temperatures by as much as 6 °C along the coast of South America. At the same time, sea surface temperatures decrease in the western equatorial Pacific and there is a reversal in the trade winds (they tend more westerly). These processes result in a decrease of the normal atmospheric pressure gradient between the subtropical high pressure cell in the eastern South Pacific and the low pressure region found to the north of Australia. At these times, below average rainfall and drought often occurs in northern and eastern Australia, subequatorial Africa, Indonesia and Chile (Chapman, 1999). Normal rainfall conditions return to both western and eastern Pacific regions when the easterly trade winds regain dominance and cool sea surface temperatures return to the west coast of South America.

The Southern Oscillation plays a vital role in promoting drought throughout substantial areas of the globe and has been estimated to be responsible for about 30% of annual global rainfall variations (Bryant, 2005). Droughts are most common during El Niño events; however, some regions experience drought during the reversed condition known as a La Niña event. During such times cold water moves northward along the South American coast. This enhances the easterly trade winds across the Pacific Ocean. As a consequence, these La Niña events can cause severe drought over the Great Plains of North America as stable high pressure cells become established resulting in a northward shift in the jet stream. At the same time, the likelihood of floods and tropical cyclones increases in eastern Australia as convection is enhanced in the western Pacific.

The occurrence of droughts has also been linked to the 11 and 22 year sunspot cycles and the 18.6 year lunar cycle. Associations between droughts and sunspot cycles are not as well defined compared to those established for the 18.6 year lunar cycle. The 18.6 year lunar cycle is defined by changes in the orbital parameters of the moon (these parameters are related to the Sun's rather than the Earth's equator). This lunar cycle affects tides which have been suggested to affect air circulation and produce reduced precipitation to, amongst other places, the lee side of the US Rocky Mountains. This supposedly affects the westerly jet stream between the Tibetan Plateau and the Rocky Mountains (Bryant, 2005).

Droughts and impacts

The Great Plains drought in the 1930s ranks as one of the greatest weather disasters in US history. The region received the name 'Dust Bowl' due to the severe dust storms that occurred during the prolonged years of the drought. At this time the Great Plains experienced severe aeolian erosion due to a lack of vegetation cover. The deep ploughing of drought tolerant native grasses in the region and replacement with agricultural plants that were not drought tolerant resulted in the exposure of soils to aeolian erosion (Abbott, 1999). The dust storms were sufficiently severe to cause dust particles to be lifted to altitudes where they were incorporated into the jet stream resulting in dust being carried as far as Europe (Robinson, 1993). The drought also caused the failure of crops and abandonment of thousands of farms.

The Sahel region of Africa has also experienced several periods of prolonged drought in recent years (Robinson, 1993). The Sahel region occurs at the southern margin of the Sahara desert where, collectively, approximately 25 million people live as farmers in the south and nomadic herders in the north. The 1968 drought was a result of a change in the position of the intertropical convergence zone, or the Equatorial Low, causing a 50% reduction in the average annual rainfall. At this time the ITCZ moved two degrees further south than normal, and exposed the Sahel region to drying winds as the precipitation that would have normally occurred here was released further south. The 1968–1973 drought was the first drought in the area to receive international attention. Between 50 000 and 250 000 people died, and more than 3.5 million cows perished. Hardest affected, however, were the children. Many were permanently retarded by disease, and cases of brain damage were related to malnutrition (Robinson, 1993).

Like the Great Plains drought, the Sahel drought was also exacerbated by human agricultural practises as natural vegetation was removed. Native shrubs and trees were cut down for fuel as people switched from kerosene to wood as their principal source of fuel. Agricultural practices also led to reduced soil structural integrity and increased surface crusting resulted in increased run-off and decreased rainfall infiltration to the soil. The environmental impacts of the drought were as severe as the humanitarian impacts. Expansion of the desert (desertification) occurred along the grasslands bordering the Sahara desert causing it to expand about 150 kilometres further south. Similar impacts were also experienced in a subsequent drought which began in 1983. At this time agricultural practices were still unsustainable and there had been a further 30–40% increase in population, resulting in even greater environmental strain (Bryant, 1991).

Droughts also affect water and air quality, animals and fish. Air quality is reduced as dust particles in the air increase. The productivity of ecosystems is severely reduced due to a lack of water which in turn results in the concentration of salts and an increase in salinity. Wildfires are also more likely during such times. All of these result in a loss of aquatic and terrestrial faunal populations and habitat due to increased vulnerability to disease and pest infestations (Chapman, 1999).

PALAEODROUGHTS

Prehistoric droughts can be identified in a wide variety of environmental records. These include:

- the mobilisation of sand dunes as recorded in the dune stratigraphy;
- changes in sediment textures and the geochemical and isotopic signatures of lake and marine sediments;
- layers of charcoal in the soils of normally humid forest landscapes;
- changes in fossil foraminifera and diatoms in marine and freshwater sediments;
- variations in pollen and palaeobotanical assemblages;
- changes in the characteristics of tree rings and annual layers in speleothems; and
- changes in isotope concentrations in fossil teeth.

The variety of signatures of prehistoric droughts is probably more diverse than any of the other natural records of extreme events. The resolution of drought palaeorecords, however, is sometimes, but certainly not always, too coarse to distinctly identify an individual drought episode that occurred hundreds or thousands of years ago. Where this occurs, such records can identify periods of time when droughts would have predominated. Other records, such as the annual records of tree rings and speleothems, can very accurately identify individual drought events that occurred centuries to thousands of years ago.

Sand dunes

Dunes are a type of aeolian or wind generated landform. They can be composed of sand grains or pellets of clay depending upon the source of the sediment. Sand dunes typically dominate arid regions of the globe today. Clay pellet dunes occur in arid regions typically around the shores of former lakes where the clay has formed into pellets on the lake floor during dry episodes. Sand dunes form when there is an abundant sediment supply, where winds are sufficiently strong to move that sediment and where there is a lack of stabilising vegetation. Their presence in humid regions today, or where they are now stabilised by vegetation, is an indication that the climate was more arid in the past (Bleckes *et al.*, 1997). Stratigraphic and chronologic investigations of dune fields, therefore, provide information on episodes of drier climates when droughts would have been more prevalent.

Periods of soil formation on the dunes have been defined as time intervals when landscape stability allowed for soil forming processes to occur. This stands to reason as the formation of soils requires the presence of vegetation which would have grown and stabilised the dunes in response to greater precipitation. Periods of landscape instability are recognised by intervals of aeolian erosion and deposition when vegetation was reduced or absent and when soils were unable to form. Increased wind velocities can cause reactivation of dunes. Once one part of a dune is activated by strong winds the moving sand can smother vegetation resulting in total reactivation of the dune. It is important to ascertain whether low precipitation, strong winds, or both, may have been responsible in the die back of the stabilising vegetation. Several studies have argued that dune reactivation is more likely to be a function of reduced vegetation cover due to a decrease in available moisture as opposed to increased wind strength (Hesse and McTainsh, 1999).

Chronologies of dune landscape stability and instability have been used to estimate past climate variations in the Great Plains, USA. There, Forman *et al.* (2001) investigated the link between past phases of dune reactivation and megadroughts. Megadroughts are suggested to last from decades to centuries and are likely to be initiated when climatic conditions cause a reduction in vegetation cover below a threshold level of ~30%. The aeolian record in the Great Plains shows that numerous megadroughts occurred during the Holocene (last 10 000 years) including several discrete events during the past 2000 years. In the central and northern Great Plains up to three events occurred in the last 1000 years and each of these was more severe than the devastating droughts of the Dust Bowl period during the 1930s to 1950s. The most recent of these appears to have occurred after the 15th Century. Figure 2.1 shows the stratigraphy and chronology of one of the dune reactivation sites in this region and highlights a period of dune stabilisation when a soil, radiocarbon dated at around 500 years BP, developed on the then dune surface. Another soil layer, higher in the stratigraphy identifies the termination of a subsequent megadrought. Such droughts, if they were to reoccur today, would bring unprecedented economic hardship to the USA.

Forman *et al.* (2001) suggest that megadroughts would have probably developed during La Niña periods when sea surface temperatures (SST) in the tropical Pacific Ocean, and later the tropical Atlantic Ocean, cooled resulting in weaker cyclogenesis in North America. Cooler SSTs in the Gulf of Mexico may have also reduced the funnelling of moisture northwards into the lower troposphere over the USA forcing the jet stream further north, producing near continent wide droughts.

Dune activity, indicative of past drier conditions, has also been noted in numerous other locations around the world including northern India, China and Australia (Bowler, 1976; Wasson, 1986; Nanson *et al.*, 1992). While many studies of dune activity can only obtain chronologies that indicate broad time spans of drier conditions, and not necessarily specific episodes of drought, these

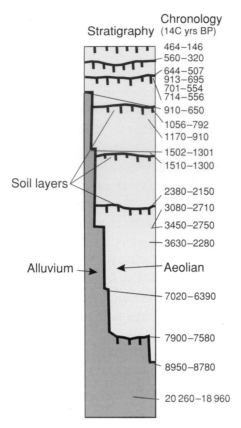

Figure 2.1. Sand dune chronology, Great Plains USA (from Forman *et al.*, 2001). The soil layers represent periods of dune stability when the climate was wetter.

studies can nonetheless highlight the rapid transitions into relatively dry phases of climate during otherwise wetter conditions. For example, Gillespie *et al.* (1983) found that the early Holocene in Ethiopia was marked by shifts into arid phases during a phase of climate which was generally becoming wetter towards the Holocene climatic optimum (HCO) (∼ 5000 years ago). The HCO was a period when temperatures and precipitation globally were generally higher than today. These marked excursions into more arid conditions occurred during the intervals 11 000–10 000, 8500–6500 and 6200–5800 years BP. The period from 11 000 years onward was generally marked by higher lake levels in Africa when precipitation is estimated to have been at least 25% higher than today (Street-Perrot and Harrison, 1984). These early Holocene conditions are often seen as an analogue of future enhanced greenhouse conditions for some regions of the globe. The Ethiopian record highlights that relatively rapid departures into prolonged episodes of severe drought are possible and may last between 10^2 and

10^3 years during phases of climate that are in the long term trending toward more humid conditions.

Rapid departures into megadroughts during the early Holocene also occurred in northern Australia. Nott *et al.* (1999) identified episodes of longitudinal sand dune reactivation near the Gulf of Carpentaria, Australia between 8000 and 6000 years BP. The chronological resolution (thermoluminescence, TL) was not sufficiently fine to determine whether this consisted of one or two episodes of drier conditions but the latter may have been possible. The dunes were totally reworked suggesting that substantial drought episodes must have occurred. There is little doubt that similar dunes would have formed in the region during the last glacial maximum (LGM), approximately 22 000 years BP, but no TL dates of this age were obtained from the present dunes despite sampling the dune cores. The early Holocene was also a period of increasingly wetter conditions in northern tropical Australia. River floods were 5–7 times larger between 8000 and 5000 years BP in the 'Top End' of the Northern Territory (Nott and Price, 1999). But in a similar fashion to tropical Africa, megadroughts in tropical Australia still occurred despite the transition towards a generally wetter climate. It would appear that there may be a global cause for this phenomenon, given its occurrence on two widely separated continents at the same time. Gillespie *et al.* (1983) suggested that the early African megadroughts might have been due to episodic volcanism, solar variability or cryospheric instability (such as glacial surges). The rapid change in climatic conditions has also been recorded in ice cores from Greenland where at approximately 8000 years BP a sudden drop in methane levels trapped in the ice suggests that the Earth plunged into a sudden chill or cold phase at this time.

Lake sediments and geochemical signatures

Layers of aeolian sediments in lakes and on the ocean floor can also be used to reconstruct past episodes of drought. Dean (1997) examined detrital clastic sediments deposited by wind in Elk Lake, northwestern Minnesota, USA. Sediments in the lake are predominantly millimetre-scale varves which are annual accumulations of sediment resulting from snow melt in the spring and lesser amounts of sediment in the winter. The spring and winter components of each varve differ in colour and sediment texture so, like tree rings, annual layers are easily identified. Aeolian sediments are also present in the lake and these are not associated with the normal varve accumulation processes (e.g. from streams). The aeolian sediments can be distinguished from the stream deposited sediments (varves) by differences in grain size. The aeolian sediments are also

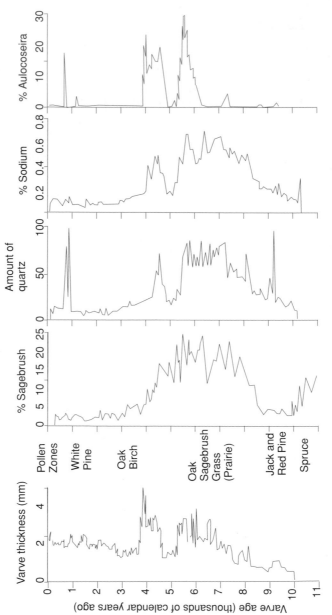

Figure 2.2. Varve thickness, pollen content and sodium levels at Elk Lake (from Dean, 1997).

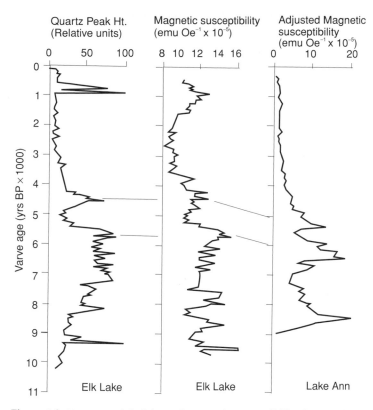

Figure 2.3. Quartz peak height and magnetic susceptibility from sediments in Elk Lake and Lake Ann, Minnesota, USA (from Dean *et al.*, 1996).

associated with several elements including aluminium, potassium, sodium, titanium and several trace elements (Fig. 2.2). The sodium (Na) can be used as a proxy for available moisture at the time of deposition because the Na is removed from the record through the decomposition of plagioclase feldspar during moister periods. The Na levels remain relatively high during the intervening dry episodes. Drier episodes, when aeolian clastic material entered the lake, are also marked by varves with higher concentrations of silt-sized quartz and increased magnetic susceptibility (Fig. 2.3).

Dean (1997) examined the 10 000 year long varve record of Elk Lake and noted that the highest concentrations of aluminium (Al) and Na occurred in the mid-Holocene (Fig. 2.2). Drought episodes then were clearly more severe than any that have occurred over the last 4000 years. Dean (1997) was also able to examine the record over the last few centuries at a finer scale because the varve sediments provide a record of annual resolution. Interestingly, the largest peaks in Na and Al concentrations during the more recent time period correspond to the Little

Ice Age (AD 1550–1700) and, in particular, the Maunder Minimum when sun spot activity underwent a prolonged (AD 1640–1710) period of quiescence. Over the past 1500 years, the aeolian record displays cyclic variations at various scales, the most obvious of which is a 400 year periodicity centred at 1200–1000, 800–600 and 400–200 years BP. These episodes are interpreted to represent episodes of windier and dustier conditions compared to the period between 3000 and 1500 year BP. The two other cycles of aeolian activity recognised in the Elk Lake varve record have periods of 1600 and 84 years. Spectral analysis shows that the former cycle accounts for 28.3% of variance in the record and the latter 14.1% of the variance. The 400 year cycle accounts for 21.5% of variance. Dean (1997) was also able to identify the episode of aridity associated with the Dust Bowl decades of the 1930s–1950s. These peaks in the record represent the greatest aeolian influx to the lake since AD 1870 and correspond to the increased regional aridity at that time. Despite the impact of this extended drought on the US economy at the time, however, the Elk Lake record shows that this phase of drier conditions was not as severe as the drought episodes during the mid-Holocene.

Holmes *et al.* (1999) also used dune and lake stratigraphies to determine past climate changes in northeastern Nigeria. The two chronologies provided a detailed and accurate picture of climate changes, from the late Glacial to the early Holocene. They found that a mainly wet phase in the late Glacial was followed by a drier period before a wetter climate dominated again between the early to mid-Holocene. Although the climate was found to be predominantly wetter during this latter period, the combined use of lake and dune stratigraphy provided sufficiently accurate records to suggest that there were smaller fluctuations in precipitation regimes. A notable environmental deterioration was noted after 4100 years BP. Since 1500 years BP the climate has been characterised by severe drought events. These findings suggest that the late 20th Century drought that has been experienced in the area may not be a unique feature, but rather a reflection of natural climate cycle.

In hydrologically closed basins, changes in lake water levels are mainly a result of changes in precipitation and evaporation. Changes in these two variables also commonly result in changes in salinity and solute concentration. These chemical changes are often recorded in the lake stratigraphy and can be analysed through sediment, geochemical and palaeontological analysis (Holmes *et al.*, 1999). The presence of greigite (Fe_3S_4), an iron sulphide mineral, in lake sediments can be indicative of past drought conditions. Greigite forms in lake sediments when lake levels are low and particularly in hydrologically open lakes during periods of low through flow of water when sulphates in the pore water of lake sediments are reduced to sulphides by heterotrophic bacteria. Reynolds *et al.* (1999) undertook a study of greigite concentrations in sediments in White

Rock Lake, Dallas, USA. They related peaks in greigite concentrations to episodes of drought identified in historical records. Authigenic greigite here has its highest peak around AD 1950, which corresponds to one of the most severe historical droughts in the region.

Calcium carbonate ($CaCO_3$) concentrations in lake sediments have also been recognised as a proxy for past climatic conditions. Calcium carbonate is regarded as a first-order proxy of climatic change because bioinduced calcite precipitation responds to changes in the timing of lake stratification and the length of the summer photosynthetic period (Mullins, 1998). Higher levels of $CaCO_3$ are deposited on the lake floor as flora and fauna respond to warmer temperatures. The reverse happens during colder and drier periods. Mullins (1998) examined the stratigraphy of Cayuga Lake in New York and found five distinct phases of cooler and drier conditions, based upon $CaCO_3$ levels, over the past 10 000 years. These dry episodes corresponded to drops in the calcium carbonate content of lake sediments. The five excursions into drier conditions can be interpreted as episodes of severe drought that lasted several centuries. Mullins (1998) suggests that the occurrence of these drier episodes followed a cycle with a period of 1800–2200 years. These oscillations are likely to reflect hemispheric and/or global climate trends, rather than local drainage-basin effects, which also correspond well with the cool and dry climate intervals recognised in the Greenland ice cores.

The strontium content of ostracods in lakes varies directly with the salinity levels of lake water. Chivas *et al.* (1985) demonstrated that ostracods buried within lake floor sediments record palaeosalinities of prehistoric lake waters and hence record changes in climate based upon the volume of water in the lake. They examined sediment cores from Lake Keilambete, Victoria, Australia which, as a volcanic maar lake, is a simple closed water basin whose water level varies directly with precipitation and evaporation. The overall amount of salt in the lake is assumed to remain constant but its concentration varies directly with the extent of evaporation of water from the system. The strontium/calcium ratio (Sr/Ca) of fossil ostracod shells over the past 10 000 years showed variations in salinity levels which corresponded well with independent assessments of past salinity levels based upon lake sediment texture. As with many other prehistoric records of lake volumes, both in Australia and globally, the Sr/Ca ratios showed reduced salinity levels and higher lake levels around 6000 years BP and a sharp increase in salinity after 4000 years BP. Salinity levels peak around 2500 years BP. Figure 2.4 shows the results of the strontium palaeosalinity measurements from ostracods in Lake Keilambete sediment cores. They clearly show fluctuations in the salinity levels after approximately 4200 years BP which could be interpreted as episodes of severe drought.

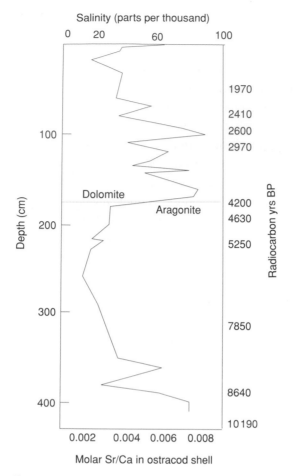

Figure 2.4. Salinity (Sr/Ca) record from ostracods in Lake Keilambete, Victoria, Australia (after Chivas *et al.*, 1985).

Marine sediments

Layers of aeolian sediment have also been recorded in sea floor sediments. De Deckker and Corrège (1991) identified layers of sand (grain size >60 μm) in a core from the Gulf of Carpentaria, Australia. They attributed these sand layers to aeolian activity during arid phases of climate. Each of the layers was deposited approximately 2250 years apart, corresponding to arid phases lasting on average 600 years and separated by wetter phases approximately 1500 years in length. The 2250 year arid cycles occurred during the Pleistocene (they did not extend their core into the Holocene) when aridity or prolonged droughts allowed the mobilisation of sand from what is now semi-arid terrain surrounding the western Gulf of Carpentaria region. The lower sea levels throughout much of the last glacial cycle caused the Gulf to exist as a lake because it was separated from neighbouring seas by shallow sills extending from the Australian mainland to Papua New Guinea.

Hesse (1994) reconstructed dust records from sediment cores from the Tasman Sea between southeast Australia and New Zealand. The resolution of the record was not sufficiently fine to identify drought episodes but it nonetheless highlights the tremendous potential for this kind of research in the future. Dust from the Australian continent is blown into the Tasman Sea by zonal westerly winds. Dust concentrations increase markedly during glacial phases of climate and highlight the onset of aridity on the Australian continent during Oxygen Isotope Stage 10 (around 400 000 years BP). Hesse and McTainsh (1999) also demonstrated from the particle grain size of the dust in the Tasman Sea cores that phases of aridity and dust mobilisation in arid and semi-arid Australia were due to a decrease in vegetation cover as opposed to increased wind strength.

Foraminifera

Foraminifera are small animals with a hard calcium carbonate body that live in marine waters. These animals have very specific environmental tolerance ranges. Species will only be found in various environmental conditions such as specific water depths, temperature ranges and salinity levels. The species *Rotalidium annectens* usually inhabits reasonably high-salinity waters and is rarely found near river mouths where salinity levels are diluted by freshwater influxes. The presence of this species in sea floor sediments near river mouths suggests that these animals were living at this location when freshwater influxes were low or absent for some period of time and salinity levels remained relatively high. Nigram *et al.* (1995) examined a core taken from sea floor sediments at 20 m water depth offshore from the Kali River in western India. They examined three aspects of the foraminifera extracted from the cored sediments; the angular asymmetry of the external test morphologies, which has been shown to be an excellent indicator of palaeo-precipitation, the percentage abundance of *R. annectens*, and the diameter of the proloculus, or the mean proloculus size (MPS). The proloculus is the initial chamber of the animal body. Figure 2.5 shows each of these parameters plotted against time (~450 years).

Nigram *et al.* (1995) found that the indicator variations in the foraminifera varied approximately every 77 years suggesting that droughts, as shown by a lack of river discharge, have been occurring with this cyclicity in this region over the past 4–5 centuries. This cycle is very similar in scale to the 80 year Gleissberg cycle of sunspot variations (Fairbridge and Fougee, 1984). Nigram *et al.* (1995) suggest that these cyclic variations in solar energy reaching the Earth

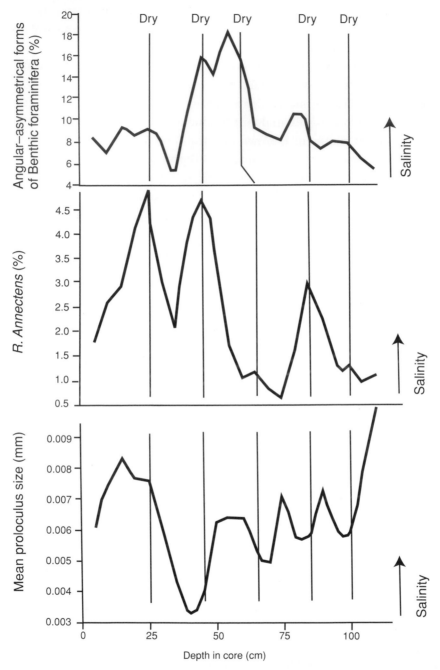

Figure 2.5. Foraminifera (mean proloculus, frequency of *R. annectens*, frequency of angular-asymmetrical forms) concentrations in cores taken offshore from Kali River in Western India. The abundance of *R. annectens* increases and the mean proloculus size decreases during droughts (from Nigram *et al.*, 1995).

may have caused the Indian monsoon to fail or occur with reduced intensity at these times resulting in higher salinity levels at the mouths of rivers in this region. Such a conclusion is not inconsistent with results from a range of other studies that also suggest the Gleissberg solar cycle may have influenced climate on Earth over the past few hundred years. Examples of this cycle have been postulated to occur in Chinese rainfall records (Hameed *et al.*, 1983; Currie and Fairbridge, 1985), the level of Nile River floods in Africa (Fairbridge, 1984), temperature variations (Agee, 1980) and atmospheric ^{14}C production (Stuiver, 1980; Stuiver and Quay, 1980).

Diatoms

Like foraminifera, diatoms can be used to infer past drought episodes. Diatoms are small silicic plants that live in both marine and freshwater environments and have species specific environmental ranges and tolerances. Gaiser *et al.* (1998) developed a model of diatom response to hydroperiods in small lakes along the US Atlantic coastal plain. Hydroperiods are defined as a measure of pond permanence and different diatoms will survive in a lake or pond depending upon the water depth and chemistry. Gaiser *et al.* (1998) showed that modern assemblages of diatoms accurately reflect the hydroperiod of lakes. Their model has not yet been applied to prehistoric systems but it promises to yield data that will accurately assess past drought episodes from fossil diatom assemblages.

Verschuren *et al.* (2000) reconstructed the drought and rainfall history of tropical east Africa over the past 1100 years based on lake-level and salinity fluctuations of Lake Naivasha, Kenya using sediment stratigraphy and the species compositions of fossil diatom and midge assemblages. They showed significantly drier climate than today during the 'Medieval Warm Period' (~AD 1000–1270) and a relatively wet climate during the 'Little Ice Age' (~AD 1270–1850), which was interrupted by three prolonged dry episodes. Figure 2.6 shows three distinct drought episodes in the early AD 1300s, between approximately AD 1450 and 1550 and around AD 1700. Interestingly, each of these drought periods occurs immediately after episodes of sunspot minima. Prior to AD 1200, drought appears to have been consistently more severe, which like the episodes of drought over the following 500 years would have placed substantial limitations on the prosperity of east African people.

Laird *et al.* (1996, 1998) also used diatoms to reconstruct the long-term hydrological history of Moon Lake, North Dakota, USA. They found a remarkable subdecadal record of lake level variations that clearly shows a dramatic change in drought regime after AD 1200. Figure 2.7 shows that prior to this time droughts

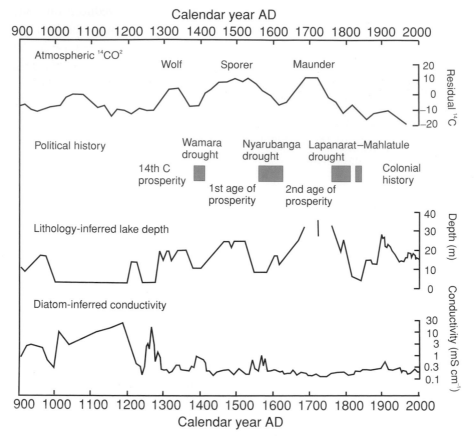

Figure 2.6. Drought and rainfall history of tropical east Africa over the past 1100 years based on lake-level and salinity fluctuations of Lake Naivasha (Kenya) (from Verschuren *et al.*, 2000).

dominated the region. After this time, few severe droughts are apparent in the record.

Charcoal layers

Charcoal layers found in sediments are often a reflection of severe bush-fires. If humans can be eliminated as the cause of these fires then it is likely that they occurred in response to drier than normal conditions. If the environment is one that is normally quite moist and rainforest or other humid climate vegetation usually exists, then the charcoal layers may be indicative of episodes of severe climatic drying and hence droughts. This could especially be the case where the environment is not normally prone to bushfires and it takes a severe

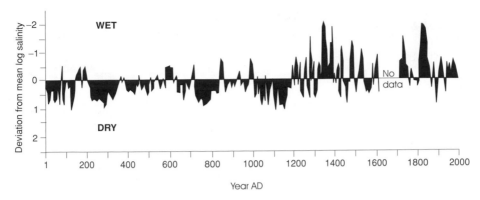

Figure 2.7. Moon Lake (North Dakota) salinity record from AD 1 to 1980 (from Laird
et al., 1996). Values below the zero line represent dry episodes.

drying of climate to desiccate the vegetation so it is susceptible to ignition.
Sediment cores derived from the rainforest floor in French Guiana suggest that
the forest here had been subject to intermittent palaeofires due to periods of
severe drought (Charles-Dominique *et al.* 1998) over the last 10 000 years. The
sedimentary and geological evidence, in combination with knowledge of prehistoric farming dates and patterns, suggest that humans were not responsible
for these fires. Archaeological evidence of human agriculture is only evident in
this area after 2000 years BP but charcoal layers exist in sediments 10 000–8000,
6000–4000 and from 2000 years BP to the present (as recent as 200 years BP). Each
of these charcoal layers occurs in the absence of any evidence for agricultural
activity.

Layers of sand and gravel are also present within otherwise clay rich organic
sediments or peats in a Pino palm swamp in the area. These coarse-grained
sediments are assumed to have been transported into the swamp by surface
water. As the swamp does not receive coarse-grained particles during modern
heavy rain events, the sand and gravel layers were probably deposited during
extreme erosional events. Charles-Dominique *et al.* (1998) argue that these events
could only have occurred following extensive deforestation due to palaeofires as
the 30–45 m high stratified vegetation canopy and dense mat of surface roots
protect the forest floor from the impacts of heavy rains and surface erosion.
Furthermore, erosion in the earlier Holocene is likely to have been less than
at present because agriculture had not yet impacted the soils. It is possible for
water deficiency to be pronounced in the South American dry season causing
fire prone conditions. Historically, however, annual humidity and precipitation
levels have been well above the water stress exceedence threshold level required
for the occurrence of fires. The record of palaeofires suggests that the area has

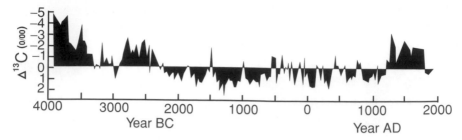

Figure 2.8. Δ^{13}C profile for peat cellulose in the Jinchuan core. Positive shifts represent lower than average soil moisture and negative shifts represent higher soil moisture or precipitation (from Hong *et al.*, 2001).

been subject to episodes of severe drought that lasted several years leading to a significant reduction of available water in the landscape.

Carbon isotopes

Stable carbon isotopes can be used to estimate the relative humidity of past climates. Different photosynthetic processes divide plants, mainly grasses, into C3 or C4 vegetation categories (Gadbury *et al.*, 2000). Plants that utilise different ways of fixating CO_2 through photosynthesis have different ratios of stable carbon isotopes such as Δ^{13}C. Variations in precipitation or soil water content are reflected in C3 plants by varying levels of Δ^{13}C. In this fashion, fossil C3 plants can be used as indicators of prehistoric changes in soil moisture and precipitation. High values of Δ^{13}C are associated with low soil moisture or precipitation, while low Δ^{13}C values are associated with high soil moisture or precipitation conditions when the plant was living (Hong *et al.*, 2001). The abundance of C4 grasses tends to be more specifically related to the minimum temperature during the growing season (Gadbury *et al.*, 2000).

Hong *et al.* (2001) used the Δ^{13}C content of peat, in Jinchuan (see Fig. 2.8), northeastern China, to estimate soil moisture or precipitation during prehistoric climatic regimes. The peat had formed mainly from *Carex* species, an annual C3 species with a shallow root system, which makes it sensitive to and dependent upon available soil moisture. Three main climate stages were identified after a radiocarbon chronology had been established. Low Δ^{13}C values, suggestive of high soil moisture or precipitation, were identified in samples aged between 4000 and 2200 BC (6000–4000 year BP). Comparisons with Δ^{18}O levels from samples within the same age bracket confirmed that this period was relatively cold and wet. These results were also in agreement with the findings from European lakes that had persistently high water levels at the same time as glaciers were advancing in many locations globally. After 4800 years BP, Δ^{13}C values showed a progressive increase suggesting that climate became gradually drier culminating in episodes of drought which have largely persisted to the present. Hong *et al.* (2001) were also able to identify eight, multi-decade to multi-century, drought periods between 2200 BC (4200 year BP) and AD 1200. At the same time Europe experienced diminished rainfall and desertification of the Saharan region was occurring. Other evidence also suggests that northeastern China experienced severe droughts over this 3000 year period. Historical records show the widespread use of astrology and meteorology to predict rainfall throughout China at the same time and prayers for rain were also recorded on bones and tortoise shells around this time. The $\Delta^{13}C$ peat record shows that a return to wet and cold conditions occurred from AD 1200 to 1800 after which dry and warm conditions have largely persisted to the present day.

Hong *et al.* (2001) were also able to use the $\Delta^{13}C$ record to demonstrate a correlation between past precipitation, temperature and solar activity. Solar activity (sunspots) changes the abundance of ^{14}C in Earth's atmosphere through variations in the amount of solar radiation reaching the Earth. Weak solar activity results in higher production of ^{14}C in Earth's atmosphere and strong solar activity lower concentrations. Hong *et al.* showed that warm and dry episodes over the past 6000 years in China correspond to strong solar activity (small $\Delta^{14}C$ values) and cold and wet periods to weaker solar activity (large $\Delta^{14}C$ values). Solar activity changed from a weak period around 2200 years BC, which was characterised by six strong peaks of $\Delta^{14}C$, to a stronger period of solar activity which corresponds to a fall in $\Delta^{13}C$ and rise in $\Delta^{18}O$ levels marking the transition into a dry warm phase. After AD 1200, solar activity weakens again and there is a corresponding fall in $\Delta^{13}C$ as climate changed back to a wet phase. Spectral analyses showed that these climatic variations correspond to periodicities ranging from 70 to 1061 years which Hong *et al.* suggest were likely to be due to perturbations in patterns of atmospheric circulation and transport of moisture principally induced by variations in solar activity.

Hong *et al.* (2003) used the $\Delta^{13}C$ record from a peat bog and *Carex* species at Hongyuan on the Tibetan Plateau to reconstruct the Indian Summer Monsoon over the past 12 000 years. The record shows abrupt variations in the monsoon over this period and reveals a clear correlation with ice rafting events of the North Atlantic (Fig. 2.9). While this record does not focus specifically on droughts or other natural hazards it does highlight how variable the monsoon has been over this time period. This of course emphasises the variability of nature over time and particularly the relative rapidity with which the system can oscillate. As the monsoon is a major controller of the frequency and intensity of floods and droughts, and tropical cyclone occurrences too, these oscillations in the monsoon highlight that the occurrence and intensity of extreme events will also be highly variable over time.

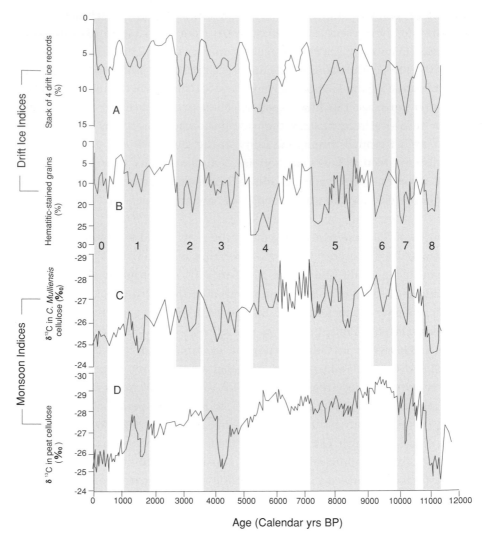

Figure 2.9. 12 000 year long $\Delta^{13}C$ record from a peat bog on the Tibetan Plateau highlighting the variability and abrupt nature of changes in the Indian Summer Monsoon and the close correlation with the record of ice rafting events in the North Atlantic. The numbers in grey columns refer to the sequence of ice rafting events starting with the Little Ice Age at 0 (from Hong *et al.*, 2003).

Oxygen isotopes

Oxygen isotopes also provide useful information about palaeoclimates. The ratio of $\Delta^{18}O/\Delta^{16}O$ in sediments, various precipitates and animal tests or skeletons can be used to estimate precipitation levels during past climates. In lacustrine environments, the oxygen isotope composition in carbonates reflects the isotopic composition of the water from which they were precipitated. The oxygen isotope composition of water bodies is usually due to changes in the origin of the water source, variations in evaporation, or a combination of the two. The oxygen isotope composition of lake waters in arid regions is largely a function of evaporation rates (Hoelzmann $et\ al.$, 2000). Because ^{16}O is lighter than ^{18}O, it is preferentially removed from lake and ocean waters during evaporation. The ^{16}O is returned to the water body by precipitation and stream runoff and these in turn are a function of the prevailing climatic regime. Hence, the ratio of ^{16}O to ^{18}O records the nature of that climatic regime. Levels of $\Delta^{18}O$ are concentrated in the lake water during periods of high evaporation (Smith $et\ al.$, 1997) when the return of water to the lake is low. As a consequence, concentrations of $\Delta^{18}O$ tend to become higher with progressive evaporation if that water body is not recharged with rainfall containing $\Delta^{16}O$. High values of $\Delta^{18}O$ in sediments and fossils suggest episodes of enhanced evaporation in comparison to rainfall. Low values of $\Delta^{18}O$ suggest that evaporation was in balance with precipitation and the body of water was not being reduced in volume. Negative values of $\Delta^{18}O$ are an indicator of times of high water input into lakes and correspondingly lower evaporation.

Water bodies that are depleted in $\Delta^{18}O$ can also indicate that the moisture that fell in this period was isotopically depleted (Hoelzmann $et\ al.$, 2000). The $\Delta^{18}O/\Delta^{16}O$ ratio can be determined by the geographical position of a site relative to the distance and nature of terrain over which rain bearing air masses have travelled. Air masses passing over continental land masses fractionate rainfall. Isotopically heavier water will precipitate first, leaving isotopically lighter water to fall farther inland. This means that unless water is added to the air masses further inland, the water that is entrained within air masses passing over continental landmasses will often be heavily depleted in $\Delta^{18}O$.

Abell $et\ al.$ (1996) examined the oxygen isotope chemistry of fossil molluscan shells in lakes of northwestern Sudan, Africa. The highly depleted $\Delta^{18}O$ values for the early Holocene suggest that the lakes were filled at this time by monsoonal rains. After approximately 5000 years BP, ^{18}O increased suggesting drier conditions ensued. The levels of $\Delta^{18}O$ fluctuated significantly in these lakes between approximately 8000 and 7000 years BP suggesting that precipitation was variable during this interval. A short sharp decrease in precipitation possibly associated with an extended drought episode occurred sometime after 7000 years BP. $\Delta^{18}O$ levels decreased markedly after 7000 years BP indicative of a pronounced wet phase after which they increased sharply suggesting a period of severe drought occurred. The return of lower $\Delta^{18}O$ levels towards the present suggests that present-day conditions are similar (wetter) to those that occurred during the early Holocene in this region.

Oxygen isotope concentrations in fossil bison molar teeth found in Nebraska, USA, suggest a serious drought may have caused the death of a large herd of these animals about 9500 years BP. The Hudson-Meng bone bed in Nebraska is a deposit of hundreds of bison that were killed by a catastrophic event. Gadbury *et al.* (2000) examined the carbon and oxygen isotopes of the bison teeth to ascertain whether an environmental signal was preserved that may give clues to the nature of this catastrophic event. One of the measures was to determine isotopic variations 'down tooth' which Gadbury *et al.* (2000) suggest is a progressive measure of the environmental conditions over the life of the animal. The bison were approximately six years of age when they died so the isotopic variations down tooth record environmental conditions over that six year period. Two of the molars examined showed down tooth variations in $\Delta^{18}O$ and $\Delta^{13}C$. The level of $\Delta^{18}O$ increases down tooth or over the life span of the ancient animal with lowermost value (being the highest value of $\Delta^{18}O$) representing the concentration in the tooth just prior to animal's death. Gadbury *et al.* (2000) regard this trend toward increasingly positive $\Delta^{18}O$ values as reflecting an increase in isotopically positive precipitation, which in turn, is a reflection of drying conditions. They suggest that the winter/early spring and/or late fall (autumn)/winter precipitation decreased with time. The isotopic concentrations also suggest a decrease in C3 vegetation over time which agrees with the interpretation of progressively drying conditions. It is possible that this progressive increase in drying of the environment resulted in a prolonged severe drought causing starvation, poor reproductive success, and finally the death of hundreds of these animals.

Oxygen isotope records have also played a significant role in unravelling the cause of the demise of the Mayan civilisation, which occupied a vast area of Central America between 2600 BC and AD 1200. They became a very prosperous people between AD 600 and 800, constructing thousands of architectural structures and were relatively advanced in astronomy and mathematics. However, their civilisation declined substantially between the years AD 800 and 900. Hodell *et al.* (1995) undertook oxygen isotope measurements on ostracods from lakes on the Yucatan Peninsula, Mexico. The $\Delta^{18}O$ measurements show distinct peaks, representing arid climatic conditions at AD 585, 862, 986, 1051 and 1391. The first peak at AD 585 coincides with the 'Maya Hiatus', which lasted between AD 530 and 630 and was marked by a sharp decline in monument carving, abandonment in some areas and social upheaval. The Mayan culture flourished from AD 600 to 800, a period shown in Figure 2.10 as a period of wetter conditions. The collapse of Classic Mayan civilisation occurred between AD 800 and 900 which corresponds to a peak in the $\Delta^{18}O$ at AD 862. At about AD 1000, mean oxygen isotope values decrease indicating a return to more humid conditions.

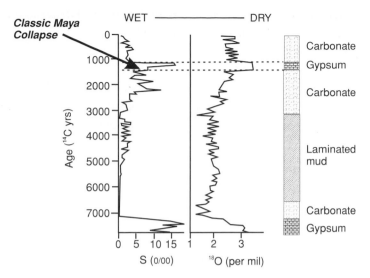

Figure 2.10. ^{18}O measurements and stratigraphy from lake sediments on Yucatan Peninsula, Mexico showing a severe drought that corresponded to the 'Classic Maya Collapse' between AD 800 and 900 (note that the figure shows radiocarbon years not calendar years; after Hodell *et al.*, 1995).

Although a post-classic resurgence occurred in the northern Yucatan, city-states in the southern lowlands remained sparsely occupied. The Δ^{18}O record suggests that drought played a significant role in causing major cultural discontinuities in Classic Maya civilisation.

Nitrogen isotopes

Based on analogues created from modern studies of fauna, nitrogen isotope analysis can be used to determine dietary and trophic relationships in fossil animals. Factors that determine the content of Δ^{15}N in collagen from herbivores include the level of soil nitrogen cycling which affects the isotopic composition of plants and the nitrogen metabolism of the animal itself. This variability in Δ^{15}N is dependent on both the animal and the surrounding environment. Factors affecting the overall level of Δ^{15}N in herbivorous animals include salinity, open environments, urea concentration, monogastric digestive systems, age and suckling (Gröcke *et al.*, 1997). Higher values of Δ^{15}N are thought to represent progressively more arid conditions.

In Australia, attempts have been made to use collagen Δ^{15}N values of macropods to estimate palaeoprecipitation levels. Analyses were conducted on species where both modern environmental and physiological characteristics were known. The results showed that changes in Δ^{15}N values correlate well with historical climate records. The technique was applied to two late Pleistocene sites in South Australia. The first site at Kangaroo Island showed higher precipitation levels compared to present, around 14 000–10 000 years BP. The second site, Dempseys Lake, also showed higher rainfall during the late Pleistocene (39 000–33 000 years BP) compared to present. While it has not been widely used as yet, this technique holds considerable promise for identifying periods of drought and higher rainfall regimes for a wide variety of regions.

Pollen and palaeobiology

Pollen is often preserved in sediments that have not experienced prolonged aerobic conditions. It can also be preserved in dung and in various carbonates such as speleothems within caves. Pollen records provide an indication of the types of vegetation previously growing in a region. Vegetation is of course a reflection of climatic conditions in any area and the preserved pollen species are indicative of the rainfall regimes prevalent at the time the pollen was released from plants and deposited.

On a global scale one can assume that vegetation regimes and their distribution are mainly dependent on climate. Vegetation is, therefore, likely to respond to climate change if the change exceeds the physiological tolerance limits of that vegetation. High vegetation mortality due to climatic factors such as frost or drought, leads to a shift in species composition. The climatic influence on vegetation is complex though. If the change in climatic conditions does not exceed the physiological limits, any response by the vegetation to that change is likely to be a result of inter-specific competition and interference. In some areas, vegetation species that are normally better adapted to the new climate, and which are found in other areas of the same climatic region, may be absent due to physical barriers to, or insufficient rates of, spread or dispersal. Although all this points to a complex vegetation response to climate change, pollen and other palaeobiological indicators are still assumed to be reliable indicators of palaeoclimates.

Pollen analysis (palynology) can also be used in combination with ecology to provide information on past climatic events (Davies, 1999). Extreme events, such as droughts and wind storms can have a marked impact upon the floristic and structural characteristics of forests. The structure of tropical rainforests for example is often composed of small juxtaposed units of vegetation that are in different stages of regeneration. Regeneration is a cyclic process starting with the fall of a large tree exposing the understorey of the forest to increased light levels. This results in the growth of pioneer species that are light demanding and short lived before they are gradually replaced by more mature forest species. The natural evolution of these units organise the species composition of the forests, and each of the individual units, around light availability. Pollen can record changes in the composition of forests which result from extreme events that result in episodic increases in light levels at the forest floor. A sudden increase in the abundance of pioneer species in a pollen record for example suggests that the forest light conditions changed rapidly which may be suggestive of the death of many mature trees during an extreme event. If the region is not prone to strong winds from tropical cyclones or tornadoes, then changes in precipitation may be inferred as a possible cause. Pollen records, therefore, can be used to identify these events and the severity of the event may be registered by the extent of the change in the record over time and area.

Van't Veer *et al.* (2000) used pollen analysis to identify the Younger Dryas in South America. The Younger Dryas was a sharp and rapid return to cooler and drier conditions around 12 000–11 000 years BP as climate was slowly warming or ameliorating after the Last Glacial Maximum. Pollen samples were taken from sediment cores extracted from the shores of several lakes. Counts of various pollen species showed a decrease in arboreal pollen around the time of the Younger Dryas suggesting a decrease in effective precipitation and temperature similar to that found in pollen species dated to the Last Glacial Maximum. The palynological interpretations were confirmed by sedimentological evidence which suggested a coincident lowering of lake levels and development of local marsh vegetation due to decreased precipitation levels.

Pollen analysis of *Hyrax* dung has also been used to reconstruct past rainfall regimes (Hubbard and Sampson, 1993). In parts of Europe, the amount and reliability of rainfall affects the relative predominance of grasses to other plants (perennial grasses versus Karoo shrublets). The content of airborne pollen reflects the composition of the vegetative ground cover and the pollen content of *Hyrax* dung reflects the pollen content in the air. Increased values of composite and decreased values for grasses indicates a reduction of rainfall and drier than normal conditions. Pollen records, however, are sensitive to local variations in precipitation, so it is important to know the seasonal mix of pellets in a fossil dung sample for this method to provide accurate estimates of past rainfall. The sample also needs to contain an even mix of seasonal markers before accurate estimates can be made. The study of modern pollen in *Hyrax* dung suggests that fossil dung could be used to provide a useful indicator of past precipitation regimes.

Problems relating to the use of fossil pollen concern the rate at which vegetation responds to climate change. Estimates have ranged from decades to millennia. The main point of argument has been whether vegetation is in equilibrium or disequilibrium with the climate. The disequilibrium model is based on the assumption that physical barriers and different spread rates cause migrational lags. Tinner and Lotter (2001), in their study of rapid changes in vegetation composition at 8200 years BP based on pollen counts from sediment cores from lakes in Switzerland and Germany, have attempted to address this issue by comparing corresponding records from the Greenland ice cores. Both the Greenland ice core and European lake chronologies indicate an abrupt climatic change around 8200 years BP (towards a cooler climate). Tinner and Lotter's (2001) pollen records indicate a substantial vegetation change for the same period, suggesting that the vegetation regime at that time, and at their study sites, was in equilibrium with climate change i.e. there was little or no lag time. Their pollen analysis was based on the presence or absence of *Corylus*, a species also found in present vegetation regimes in Central Europe. The period around 8200 years BP was also found to coincide with the appearance of previously absent or rare taxa. Tinner and Lotter estimated the vegetational lag in response to climate change in this case to be no more than a few decades.

Another issue arises from the Tinner and Lotter (2001) study. Their results indicate that it would be easy to suggest that *Corylus* is affected (i.e. decreases in abundance) by colder temperatures and that climatic change was the specific factor behind the vegetation changes that occurred simultaneously with a significantly cooler climate. This is unlikely, however, as *Corylus* is presently found within the climatic range estimated for this cooler period around 8200 years BP. The decrease of *Corylus* is more likely to be the result of inter-specific competition. Dry weather conditions and drought are usually accompanied by a reduction of the radial and height growth of vegetation along with foliage loss. This causes tree crowns to thin resulting in more light reaching the under-storey of the forest. Tinner and Lotter (2001) have hypothesised that drought was common prior to 8200 years BP which favoured the light dependent *Corylus*. When drought stress decreased as a result of cooler temperatures, taller growing trees were favoured and out-competed *Corylus*, because it is a shade-intolerant species. These findings emphasise the complexity of using pollen records and other biological data to estimate past climate change.

It is not always possible to rule out human interference in vegetation composition when pollen analyses are used to estimate prehistoric precipitation regimes. If human interference has occurred, the pollen analysis will not describe a natural chronology. Although this may pose a problem in some studies, it has been assumed that such interference would result in a continuously decreasing trend in species composition rather than a cyclic trend. Cyclic trends that range over longer time spans are therefore often assumed to be natural chronologies (Bonnefile and Chalie, 2000).

Fossil leaves have also been used to indicate past rainfall. This is based upon the observation that the morphology of a living leaf, and especially the size of the leaf, is influenced by the available moisture. Plants found in more arid areas tend to have smaller leaf size than plants found in more humid regions. This occurs as plants transpire water into the atmosphere through the surface area of the leaf, thus a larger leaf has greater surface area and consequently loses more water (Wilf *et al.*, 1998).

Estimates of palaeoprecipitation based on leaf morphology are often based on a method called Wolfe's Climate Leaf-Analysis Multivariate Program (CLAMP). This method ordinates a multivariate data set of leaf-morphology characteristics scored from modern vegetation samples that are associated with climate stations. In this way, a quantitative framework can be developed for estimates of climatic variables. At present, most data of this kind is based on vegetation from North American forests and there is a profound lack of data from tropical regions. The technique has been found to overestimate precipitation, however, and such tendencies have to be considered when palaeoclimates are reconstructed. The overestimation is larger when data is extrapolated from the original data in North America to other vegetation communities in warmer climates. Due to these issues with the CLAMP method, Wilf *et al.* (1998) investigated a univariate method of climate reconstruction based on leaf-area analysis. This method produced more significant results than the multivariate models based on the CLAMP data set. Although the univariate method has proved to be more suitable than the multivariate method, it is advisable that additional data should be used where possible.

Tree-ring analysis (dendrochronology)

Tree-ring analysis can be used to indicate precipitation levels for historic or prehistoric times on temporal scales of decades to millennia (Hughes *et al.*, 1994). Most commonly, the method is used to estimate precipitation levels ranging over the scale of centuries. The use of tree-ring analysis is particularly applicable to temperate and subpolar regions. Here, several long-lived species have been found to leave identifiable traces of past precipitation events (Loagicia *et al.*, 1993).

Dendrohydrology is the study of long-term hydrological variability using tree rings. The technique is dependent upon a prior understanding of the relationship between hydrological variables and the biological response of trees to these variables. The hydrological variables include precipitation, run-off, near surface temperature, evapotranspiration, ground water and soil moisture. The response of a tree to changes in these variables is recorded mainly in the width of

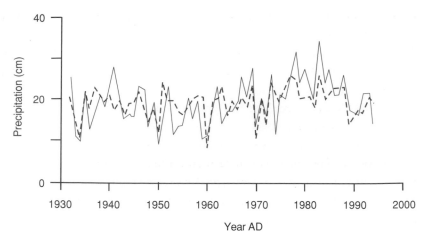

Figure 2.11. Comparison between the instrumented precipitation record and the tree-ring record, California, USA. The solid line represents the instrumented record and the dashed line represents the reconstructed (tree-ring) record (after Hughes and Graumlich, 1996).

individual tree rings. However, other biological and environmental conditions can affect inter-annual fluctuations in ring width, and current ring growth. Because of this, ecological conditions such as soil fertility, plant competition, forest stand development, diseases and insect pests, wildfire incidents, environmental pollution and sunlight availability that affect tree growth must also be considered when undertaking dendrohydrological studies (Loagicia *et al.*, 1993).

Tree-ring analysis is usually better suited for the reconstruction of past drought events rather than past events of high precipitation. This is because conditions such as plant competition and soil fertility act as limiting factors in years with high precipitation and obscure the hydro-climatic signal in the tree-ring records. Lag effects after severe conditions can also obscure tree-ring analysis as individual events can affect forced responses in the tree-ring width for several years. Despite these issues, however, tree-ring analysis can provide accurate descriptions of past precipitation regimes. Figure 2.11 shows a comparison between the instrumented precipitation record and a tree-ring record in the USA. The tree-ring record follows the general trend of precipitation variations; however, in places there are slight differences in the magnitude of events or seasonal rainfall. Despite the departures in magnitude, the comparison shows that tree rings can provide a true reflection of precipitation changes. The main limitations preventing tree-ring analysis from becoming a more commonly used tool to create palaeoclimate chronologies are a lack of long-living, climate-sensitive trees in many parts of the world (such as Australia), the elaborate process of field sampling, laboratory and statistical analysis necessary to produce accurate estimations and the limited exposure of mainstream hydrologists to the field of dendrohydrology (Loaiciga *et al.*, 1993).

Woodhouse and Overpeck (1998) analysed tree-ring data from the Great Plains region in the central USA, to gain an accurate record of droughts extending back to about AD 1300. They found that 20th Century droughts are not representative of the full range of droughts that have occurred in the USA over this period and probably the past 2000 years. Multidecadal droughts of the late 13th and 16th Centuries were more prolonged and severe than those of the 20th Century and few, if any, major droughts have occurred in the periods between these episodes of severe drought (see also Stahle *et al.*, 2000). The long-term record also suggests that there has been a regime shift in droughts after the 13th Century (Stine, 1994). Droughts of the 20th Century have been characterised by moderate severity and comparatively short duration. Droughts prior to this time were at least decades in duration, whereas after this time (i.e. 20th Century) droughts were a decade or less in duration. The records also highlight that there can be substantial periods of time between episodes of truly severe drought. Woodhouse and Overpeck (1998) stress that the same is true of lake sediment, microfossil and isotope histories of droughts from the USA, Australia and Africa. Such records provide a warning that the short historical records of droughts are often not a reliable guide to drought climatology and 20th Century records are less than ideal as a guide to future events.

Tree-ring records of droughts have also played a role in unravelling some intriguing mysteries concerning the demise of some human populations. The first group of English colonists to the USA landed on Roanoke Island, located off the NE coast of North Carolina in AD 1585. They did not last long and returned to England a year later. A second group of colonists, arriving in 1587, mysteriously disappeared and not a single colonist was found by the time additional supplies were brought from England in 1591. The ring record of baldcypress trees (*Taxodium distichum*) was used by Stahle *et al.* (1998) to reconstruct the climatic history for this region. The record suggests that the most severe drought over the past 800 years coincided with the disappearance of the Roanoke Island colonists (Fig. 2.12). Stahle *et al.* (1998) also showed that a severe seven year drought (AD 1606–1612) in Jamestown, the main settlement, may have been the cause of a high death rate in the colony around this time. Only 38 of the original 104 colonists survived the first year (AD 1607) of the new settlement at Jamestown

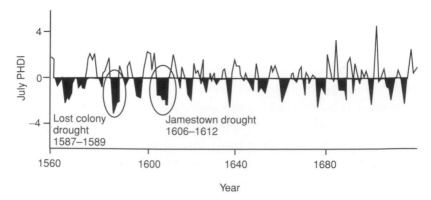

Figure 2.12. Tree-ring reconstruction of droughts using the July Palmer hydrological drought index (PHDI) showing major droughts that affected the first colonists in the USA (from Stahle *et al.*, 1998).

and only 3400 people survived of the original 6000 who settled there between AD 1608 and 1624. Most people died of malnutrition.

Speleothem records

Speleothems are deposits of calcium carbonate often found in limestone caves. They can house information of past climates over a greater temporal span than that often achieved accurately through tree-ring analysis (Ming *et al.*, 1997). Like trees, speleothems can produce annual banding. The banding is usually best developed in speleothems in temperate climates with strong seasonality. Banding can be expressed by a number of characteristics such as alternating fluorescent and non-fluorescent calcite layers, alternating fibrous and less dense fibrous layers, and alternating aragonite/calcite layers. Fluorescent (luminescent) banding develops where strong seasonal changes in temperature result in marked decreases in soil microbial activity and organic acid solubility during the winter. Porosity/density banding is due to marked seasonality in precipitation where growth rates and fluid saturation are controlled by the precipitation regime. Mineralogically defined banding (alternating aragonite and calcite layers) also reflect marked seasonal changes in precipitation patterns (Liu *et al.* 1997).

Liu *et al.* (1997) examined a stalagmite (speleothem extending upwards from a cave floor) from Shihua Cave, Beijing and identified three kinds of annual banding; a luminescent banding that is only visible under a fluoromicroscope; a lamina layer which is visible to the naked eye; and a transparent microband that can be observed through a polarising microscope. The transparent microbanding

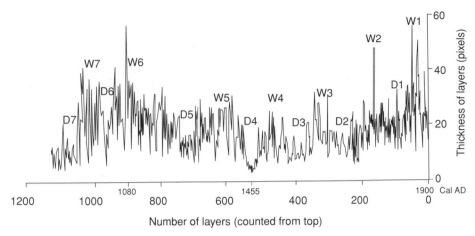

Figure 2.13. Annual record of rainfall and drought, Shihua cave, Beijing, China from a speleothem over the past 1100 years (from Liu *et al.*, 1997). D = drought and W = wet.

consists of a light coloured zone at the base of the band, which is likely to have been deposited by old water, and a dark part at the top of the band deposited by fresh water. They developed a time series comparing individual layer thickness with an historical index of drought and flood (Fig. 2.13) and the instrumented record of precipitation from AD 1951 to 1980. Both records agreed on the years of high rainfall providing confidence in the interpretation of the older part of the stalagmite banding record. The thickness of individual layers was taken as a proxy for the level of annual precipitation. They found several peaks in annual rainfall and years of drought over the past 1100 years. Seven episodes of drought were identified and spectral analyses highlighted several climatic cycles with periodicities of 136, 50, 16–18, 11 and 5.8 years. There was also evidence for possible millennial scale variability.

Conclusion

Palaeodrought records provide one of the more accurate and informative archives of the long-term nature of an extreme natural event. Annual layers in tree rings, speleothems and lake sediments show that droughts, both in terms of their severity and duration, can be highly variable over time. One of the clear messages from these records is that 20th Century droughts are a poor reflection of the nature of this hazard in previous centuries. Droughts of the late 13th and 16th Centuries in the USA for example were more prolonged and severe than those of the 20th Century and Dean (1997) has recognised that droughts here, over the long term, display 400 year periodicities. These aspects of the drought regime cannot be recognised from short historical records alone. Indeed, to assume that the historical record is a realistic reflection of the nature of this hazard would clearly lead to a gross underestimate of community vulnerability and exposure. The long-term drought records highlight the importance of complementing short historical records with long-term natural archives to gain an understanding of the possible impacts of this hazard in the future.

3

Floods

Floods occur when water from terrestrial run-off leaves a stream channel and spreads across the surrounding landscape. The definition of a flood can take on different meanings to different people. Many would consider flooding to occur only when the level of water has risen to a point where the threat to property and infrastructure is unavoidable (Bell, 1999; Chapman, 1999; Smith, 2001; Bryant, 2005). In this sense flooding is viewed as a hazard. Floods are also the maintainers of ecosystems and support life in coastal estuaries, lakes, wetlands and enrich vast floodplains and play an integral role in the geomorphic evolution of landscapes (Jones, 2002). Flooding is a natural function of river behaviour and floodplains by definition are innately flood-prone (Chapman, 1999).

Causes of floods

The causes of floods can be broadly divided into physical, such as climatological forces, and human influences such as vegetation clearing and urban development. The latter can result in an exacerbation of the conditions affecting the run-off of water resulting in flood intensification (Smith, 2001). The most common causes of floods are intense and/or prolonged storm precipitation, rainfall over areas of snow cover, rapid snow melt, the successive occurrence of medium to major size storms and the failure of dams, including ice dams (Chapman, 1999). The most important causes of floods are atmospheric hazards, most notably rainfall. Floods can also be associated with oceanic and atmospheric processes on a large scale such as the El Niño Southern Oscillation (ENSO) phenomenon. For example, the 1993 USA floods were associated with an ENSO event and the 1988 flooding in Sudan and Bangladesh were linked to a La Niña episode (Smith, 2001). The floods of eastern Australia in 1974–75 also followed a major El Niño event, as the Walker Circulation intensified resulting in a La Niña event bringing two years of the heaviest rain in recorded history to the region (Bryant, 1997).

Meltwater from snow and ice is also a source of floods. Meltwater flooding is a problem in areas draining mountains and places with significant winter snow accumulation (Bryant, 1997). Melting snow is responsible for flooding in the late spring and early summer on the continental interiors of Asia, North America and some parts of Eastern Europe (Smith, 2001). Some of the most threatening events are from rainfall on snow combined with rapid spring warming resulting in large water flows in a short period of time (Smith, 2001; Bryant, 2005). Meltwater floods can also be compounded by ice jam flooding, which occurs through an accumulation of large pieces of floating ice that build up in the river system, often at bridges, resulting in a damming effect (Smith, 2001). Large ice masses in snowmelt can damage buildings, destroy trees and dislodge houses (Smith, 2001). The Romanian floods of 1970 were a result of heavy rain from a deep atmospheric depression during annual snowmelt from the Carpathian Mountains (Smith, 2001).

Flash floods are short-lived extreme events. They most commonly occur under slow moving or stationary thunderstorms when rainfall exceeds the infiltration capacity of an area resulting in rapid run-off (Bell, 1999). They tend to last less than 24 hours and are destructive due to the high-energy flow. Flash floods are particularly hazardous due to the quick onset of floodwaters and the lack of time for adequate warning systems to be activated (Blaikie *et al.*, 1994). Flash flooding in semi-arid and urban areas can be more destructive because extensive sheet flow can occur within very short distances due to a lack of vegetation to defuse raindrop intensity (Bryant, 2005). This results in a high run-off to infiltration ratio and as a consequence the water can enter the channel quickly resulting in a high, rapid onset, flood peak.

Prolonged rainfall events are the most common cause of flooding worldwide. These events are usually associated with several days, weeks or months of continuous rainfall. Prolonged rainfall over large drainage basins can be associated with monsoonal activity, tropical cyclones or intense depressions of the mid-latitudes (Smith, 2001). Heavy rain can form in the large vertical development of storm clouds with strong updraughts associated with large-scale low-pressure systems, or as a result of large volumes of atmospheric moisture sucked into convection cells or orographic uplift along coastal mountain ranges (Jones, 2002). The longer the duration of the rainfall event the greater the amount of surface run-off entering a river channel (ARMCANZ, 2000). A prolonged rainfall event in 1861 in Cherrapunji, India, was caused by monsoonal winds carrying unstable moist air from the Bay of Bengal up over the Himalayas. This system produced

9.3 m of rain for July and contributed to an astonishing yearly total of 22.99 m that resulted in widespread flooding (Jones, 2002).

Human-induced floods

Human impacts on river catchments influence flood behaviour. Land use changes in particular have a direct impact on the magnitude and behaviour of floods (Chapman, 1999). Deforestation results in increased run-off and often a decrease in channel capacity due to increased sedimentation rates (Smith, 2001). A four-fold increase in flood peak has been measured in some smaller sized catchments following forest clearance (Smith, 2001). The 1966 floods in Florence, Italy, which claimed 33 lives, were at least partly due to long-term deforestation in the upper Aron basin (Smith, 2001). Bryant (2005) suggests that flash flooding in urban areas has increased in recent decades due to the increase in deforested land and the ever-increasing expansion of the urban sprawl with its impervious surfaces and inadequate drainage systems.

Smith (2001) suggests four major ways in which urbanisation impacts on floods including the concentration of highly impermeable surfaces, insufficient urban storm water drainage networks, channelisation works and constricted channels. Surfaces such as roads, roofs and concrete areas inhibit infiltration and increase the speed at which run-off travels (Bell, 1999; Chapman, 1999). Channelisation works on a river can reduce the carrying capacity of a river system; navigation works on the Mississippi River in the USA, for example, have reduced the capacity of the channel by one-third (Bell, 1999). Insufficient drainage of storm water following urban development also contributes to urban flooding. This can be directly related to the design capacity of urban drains which, as Smith (2001) notes, is often for storms with a return interval of 1 in 10 to 1 in 20 years. In tropical countries, up to 90% of lives lost through drowning from flooding are a result of precipitation on small steep catchments with poorly drained urban areas (Smith and Ward, 1998).

Dam breaks can result in floods of truly monumental size. There are many different reasons why dams break or fail. These include a sunny day failure of the main dam wall which is an unexpected failure of the dam not associated with flooding or natural disaster, and high-intensity rainfall events and earthquakes that can crack the foundation of the dam. The most common reason dams fail is because of flooding resulting from probable maximum precipitation events in a catchment area. The failure of dams due to precipitation events has been attributed in part to inadequate design for high-magnitude events combined with geological position and the general age and neglect of the dam (Bryant, 2005). The largest death tolls from flash floods have been associated with the failure of dams. In 1963, the Vaiont Dam in Italy failed killing 2000 people and in 1889, 2200 people were killed in Johnstown, Pennsylvania from a dam failure flood (Bryant, 2005).

Characteristics of flood flows

Rivers are an important part of the hydrological cycle. They are a conduit for precipitation generated run-off where the water is transported to lakes, swamps, dams and/or oceans. The extent of a rain-induced flood is dependent upon the alignment of the storm to the basin, the rate of movement of the storm and the spatial extent of the storm (Chapman, 1999). The passage of floodwaters is governed by the configuration of the catchment including its shape, area, slope and drainage density (Jones, 2002). The hydrological and biological characteristics of the system are also important factors (Chapman, 1999). Catchment area is the most important factor in controlling the volume of discharge (Jones, 2002); the larger the catchment, the longer it takes for the total flood flow to pass a given point (Bell, 1999; Bryant, 1997). Also, the size of a catchment has a bearing on the amount of precipitation it can receive. Larger catchments, for example, can receive a considerably greater volume of rain than a small catchment in a storm of large spatial extent.

The run-off to infiltration ratio of a catchment is another important factor governing the behaviour and size of floods. Run-off is the surface water that remains after evapotranspiration and infiltration into the ground and comprises water from surface sheet or channel flow; it is the major component of the flood peak discharge during a precipitation event (Bell, 1999) (Fig. 3.1). The total run-off from a catchment is made up of direct precipitation into the stream channels, surface run-off, interflow and base flow. The larger the response and travel time (in terms of velocity) of surface run-off, the bigger the build up of peak flow. The slower the run-off travel time, the greater the chance of that surface water infiltrating the ground surface. This water will then take longer to reach stream channels resulting in a more subdued flood peak. If the catchment contains a large number of lakes or swamps, run-off rates are likely to be reduced as they tend to absorb high peaks of surface run-off (Bell, 1999). High slope gradients and drainage densities contribute to the speed of run-off; the shorter the length of the hillslope, the faster that run-off will reach and contribute to flood flow (Jones, 2002).

Measurement of flood flows

The peak discharge in a river is responsible for the maximum inundation of an area. The discharge rate is the basis for most methods involved

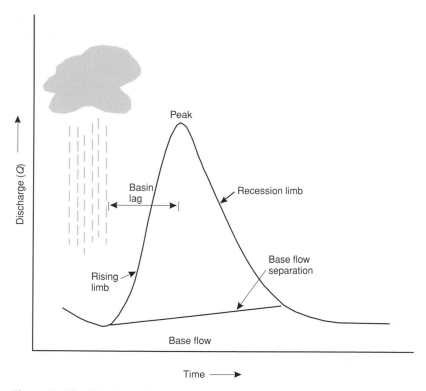

Figure 3.1. Flood hydrograph.

in predicting flood magnitude (Bell, 1999). The peak discharge of a flood can be used to determine the depth and area of the flood as well as the duration and velocity of the flood waters (Bell, 1999). The probability of a flood event occurring is determined by fitting a frequency distribution to a sample of flood observations from a given stream. This helps to determine the recurrence interval of a particular magnitude event (Chapman, 1999). To measure the amount of damage potential a flood may have it is necessary to consider the stream power of an event and the amount of energy developed per unit time along a river boundary (Bryant, 2005). The largest calculated stream power was from the late Pleistocene Missoula floods in the USA. These floods reached velocities of 30 m s^{-1} and depths of 175 m with a stream power in excess of 1 000 000 W m^{-2} (Bryant, 2005).

Flood flows in river systems can be analysed via the discharge hydrograph. The hydrograph is a curve describing the volume of discharge with time at a point along the stream channel (Fig. 3.1). A hydrograph can be used to determine the total flow, base flow and periods of high and low flow showing yearly, monthly, daily or instantaneous discharges (Bell, 1999). Hydrographs are useful in predicting the passage of a flood through a river system from a single storm event (Bell, 1999). The rising limb of the hydrograph concaves upward reflecting the infiltration capacity of the catchment (Fig. 3.1). A sharp rise in the hydrograph curve occurs as the infiltration capacity is reached and there is a sudden surge of surface run-off (Bell, 1999). This is followed by a peak in the curve that marks the maximum run-off from an event and occurs at some defined time following the rainfall peak (Chapman, 1999). This period of time between the rainfall peak and flow peak is referred to as the lag time (basin lag) of the event. It is possible that some basins may have two or more peaks in the hydrograph curve for a single event depending upon the nature of the rainfall event and the catchment characteristics (Bell, 1999). The downward limb of the curve represents the reduction of water into the system from surface run-off and precipitation compared to the delayed delivery from interflow pathways, regolith and groundwater storage (Chapman, 1999). Again, the slope of the recession limb is dependent on the physical characteristics of the catchment (Bell, 1999).

Floods as a natural hazard

A flood event is not considered to be a natural hazard unless there is a threat to human life and/or property (Smith, 2001). The most vulnerable landscapes for floods are low-lying parts of floodplains, low-lying coasts and deltas, small basins subject to flash floods, areas below unsafe or inadequate dams, low-lying inland shorelines and alluvial fans (Smith, 2001). Rivers offer human populations transport links, a water source, recreational amenities, flat often fertile plains and are an attractive place for settlements. Floods then become a major natural hazard because of the high human population densities that inhabit these lands.

The direct impacts of a flood are closely related to the depth of inundation of flood waters. The extent of a flood has a direct relationship for the recovery times of crops, pastures and for the social and economical dislocation impact to populations (Chapman, 1999). In urban areas, the level of flood damage is related to the type of land use, the depth of water and period of inundation (ARMCANZ, 2000). In rural areas, the damage caused by floods is also dependent upon the type and growth stage of the crops at the time of flooding. In addition to the loss of crops and livestock, rural areas also suffer loss of topsoil and fertilisers, with extensive damage to fences and increased weed infestation (ARMCANZ, 2000). Sediment loads, floating debris such as trees, pathogens and pollutants in floodwaters also have an impact. The damage potential of floods increases exponentially with flood velocity (Smith, 2001).

Social and economic impacts of floods

Floods are responsible for up to 50 000 deaths and adversely affect some 75 million people, on average, worldwide every year (Smith, 2001). Drowning is not the only cause of death in floods. Disease is common after floodwaters pass, especially in less developed countries. Gastrointestinal diseases mainly break out due to damaged sewerage systems and low sanitation standards. Malaria and typhoid outbreaks after floods in tropical countries are also common. In many western countries it has been observed that survivors of catastrophic flood events suffer some form of mental illness that can be directly related to the flood event. French and Holt (1989) concluded that flood victims can suffer psychological problems for up to five years after the flood event. The emotional strain of the event on financial and social costs would appear to be a catalyst for these impacts (ARMCANZ, 2000). The Buffalo Creek, West Virginia floods of 1972, left 90% of survivors with mental anguish 18 months after the event (Smith, 2001).

The floodplains of the Yangtze River are home to 5% of the world's population (Chapman, 1999). Five million people lost their lives in floods in China between 1860 and 1960. The Chinese have had flood defence for cities for the last 4000 years but large losses still continue with 3000 lives lost and 15 million people left homeless in the 1998 floods (Smith, 2001). Deaths from flooding in Bangladesh account for nearly three-quarters of the global loss of life from floods (Smith, 2001). It has been estimated that in India and Bangladesh 300 million people live in areas that are affected by floods. Chapman (1999) suggests that 3000 people and 100 000 head of cattle perish in flooded river systems in these countries every year. Indeed, in 1991, 140 000 people were killed in floods in Bangladesh. In 1993 in the US midwest, the Mississippi and Missouri Rivers affected nine states, with 50 000 damaged homes resulting in the evacuation of 54 000 people from flood affected areas (Smith, 2001).

Physical damage to property is one of the major causes of tangible loss in floods. This includes the costs of damage to goods and possessions, loss of income or services in the flood aftermath and clean-up costs (ARMCANZ, 2000). There are also secondary losses such as declining house values where properties are deemed to be in a flood prone area. Some impacts of floods are intangible and are hard to place a monetary figure on. For example, every year in Bangladesh it has been estimated that riverbank erosion of farmland leaves one million people landless (Smith, 2001). Intangible losses also include increased levels of physical, emotional and psychological health problems suffered by flood-affected people (ARMCANZ, 2000).

Floods are an expensive phenomenon, with the estimated average annual cost of flooding in Australia alone being AUS $350 million (ARMCANZ, 2000). In 1948 a major snow melt flood from a rapid temperature rise after heavy winter snow in British Columbia, Canada, resulted in more than 2000 people left homeless and a compensation bill in excess of Can $20 million (Smith, 2001). The 1998 Chinese floods left a damage bill of US $20 billion (Smith, 2001). The 1993 Mississippi River floods resulted in a damage bill between US $15–20 billion and saw a 17% reduction in the national soybean and corn yields (Chapman, 1999). In India, 75% of the direct flood damage is attributed to crop loss (Smith, 2001).

Floods, therefore, are one of the most costly and wide reaching of all natural hazards. Our ability to model the size and extent of a flood during a heavy rainfall event has improved greatly in recent years. Because of this, emergency managers and hydrologists can estimate with a reasonable degree of accuracy the area of land likely to be inundated if detailed digital terrain models and numerical flood models are available. Many lives can be saved in these situations through effective and timely evacuations. But property damage and agricultural losses can still be high. In many ways our ability to mitigate against floods is advanced. But as with many hazards, and despite our technological tools, many urban developments are still allowed to occur in flood prone areas. Palaeorecords, while often ignored, can provide us with vital information on the magnitude and frequency of the flood hazard and also the extent of a landscape likely to be inundated in the future. Many risk assessors are unfamiliar with these long-term records and the methods by which they can be uncovered. As a consequence they limit their understanding of the flood hazard by restricting their analyses to the historical record.

PALAEOFLOODS

A variety of natural phenomena record floods. These include:

- slackwater sediments;
- plunge pool deposits;
- flood transported debris such as tree branches and logs left high on river banks or valley sides;
- deposits of large boulders in river channels and the spacing of boulder bars and rapids along the length of a river; and
- the shape and form of a stream, otherwise known as channel geometry.

Each of these either record the height of the water during a flood or is a measure of the magnitude of the flood flow and its competence to move objects of a certain size and density. Some forms of palaeoflood evidence are a direct measure of the flood height (usually minimum height) and in this sense are not strictly proxies but an actual record of that event. Other forms of evidence, such as those that measure flow competence, can be regarded as proxy measures of flood height.

The development of the discipline of palaeoflood hydrology is a classic story of the reluctance by the scientific community to accept new ideas. J. Harlen Bretz was the first to argue that the deposits and erosional features present in the Missoula scabland in Washington State were produced by truly massive flood flows. These features formed when water was released after ice dammed lakes burst during the amelioration of the climate following the Last Glacial Maximum (Bretz, 1928). Bretz was attacked for his views that these features, which are phenomenally large compared to most fluvially generated forms, could have been caused by possibly the largest ever floods on Earth in recent geological time. He was later shown to be correct in his interpretation and as a result of his early research, and then the subsequent research of many others (principally by V. Baker of the University of Arizona and colleagues), the science of palaeohydrology emerged. Today, palaeohydrology is seen as most credible and a very valuable tool in the understanding of flood climatology for regions globally. However, while the science is now well established, its incorporation into flood risk assessments has not been so readily adopted everywhere.

Slackwater sediments

Slackwater sediments are generally fine-grained sediments deposited in zones of low velocity flow during a flood. Such zones of low flow velocity include the mouths of tributaries (Fig. 3.2), alcoves and caves in bedrock gorge walls, areas where channels expand in width, upstream of channel constrictions and the sides of river valleys where sedimentary terraces and levees can develop (Fig. 3.3). Each of these zones represents an area separated from the main flood flow where circulation eddies or physical obstructions cause the flood flow velocity to slow and as such deposit the suspended sediment. The fine-grained size of these sediments shows that the flow at the point of deposition must have been of relatively low velocity. As a general rule, there is a direct relationship between the grain size of the sediments deposited and the local flow velocity such that progressively finer-grained sediments are deposited with decreasing flow velocity. The height of the sediments above the floor or bed of the river provides an indication of the stage height of the flood flow. This stage height is usually only a minimum flood flow height as the real flood water height is at least 20% higher than the elevation of the slackwater sediments.

The height of a number of slackwater deposits along a river reach, often a length of gorge, can be used to model the discharge of the palaeoflood event. The modelling is usually undertaken using a step backwater approach that calculates

Figure 3.2. Katherine Gorge, Northern Territory, Australia. The tributary mouth in the left foreground is a typical location for slackwater sedimentation. Flow is from background to foreground.

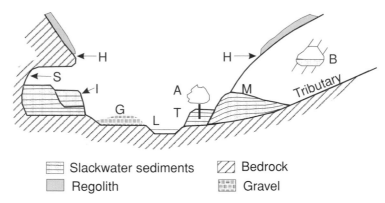

Figure 3.3. Schematic diagram of typical slackwater sediment settings and palaeostage indicators (after Kochel and Baker, 1988). H, scour of soil or regolith; S, silt line; B, cave deposit; M, mounded slackwater deposit at tributary mouth; T, slackwater terrace; I, inset slackwater deposit; A, flood damaged tree; G, gravel bar; L, low water flow in channel.

the energy grade line or slope of the floodwater surface. The grade line can then be used to calculate flow velocity, discharge and stream power. The age of the flood is usually determined by using one of several geological dating techniques (radiocarbon or luminescence, see Appendix A) on the sediments or organic inclusions within the sedimentary unit. Often slackwater sediments accumulate

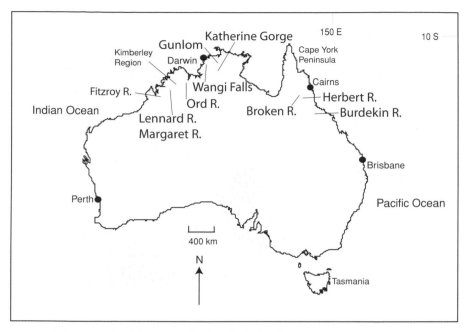

Figure 3.4. Location map of palaeoflood sites in northern Australia.

into a sedimentary stack where multiple, individual flood events each leave a sedimentary layer. The law of superimposition applies in such circumstances so the uppermost layer is the youngest and each successively lower layer becomes progressively older. When dating the layers, the accuracy of the technique used can be partly tested by the stratigraphic integrity of the dates i.e. each layer should return a progressively younger age with upwards stratigraphic order. Layers are often separated from each other by a thin layer of leaves and other organic debris and/or a drape of clay. Gillieson *et al.* (1991) found four such flood stacked slackwater deposits in alcoves of the walls of the limestone Windjana Gorge (Lennard River) in the west Kimberley region of northwest Western Australia (Fig. 3.4). The age of the layers was determined by radiocarbon dating freshwater mussel shells at the base of layer 2 and thermoluminescence dating of quartz sand within other layers. The four flood events occurred between AD 1962 and 2800 years BP.

While many slackwater deposits are composed of stacked relatively thin (<1 m) layers, other deposits can be made up of one very thick layer possibly representing a single very large flood event. One such sequence occurs in the Ord River Gorge of northwest Western Australia. The gorge is about 20–50 m deep and is carved into Proterozoic sandstones. It does not maintain a uniform width along its length; in places the width decreases from about 300 to 200 m.

Figure 3.5. Major slackwater sediment terrace (arrowed) in Ord River Gorge, Western Australia. Note the trees in the foreground in comparison to the height of the terrace.

Sedimentary terraces occur immediately upstream of these gorge constrictions in at least five separate locations. The terraces stand up 12 m high above the normal dry season river level and extend along stream for several hundred metres (Figs. 3.5 and 3.6). They are flat topped and usually about 30–40 m wide and abut the gorge walls. They can be likened in morphology to a lateral bar. They are composed of well-sorted, fine-grained sands and silts and show no distinct variation in colour, texture or degree of weathering down profile. Indeed for the most part, the sediments have a relatively fresh appearance as they contain well-preserved micas, which are normally easily weathered and removed from sedimentary units in tropical Australia after only a few thousand years. Sedimentary structures are also preserved within these units. These 'drape-like' structures appear to be bedding planes developed from the deposition of suspended sediments in relatively still water for they have high dip angles and drape over irregularities along the contact with the underlying strata. Optically stimulated luminescence ages from two of these terraces, about 2 km apart, suggest that the sediments were deposited between AD 1300 and 1800. It is uncertain at this stage whether the terraces were deposited during a single flood event or a series of closely spaced events over this period. The height of the terraces, and hence

Figure 3.6. Close up of a slackwater sediment terrace in Ord River Gorge. Note the person (arrowed) in the right foreground for scale.

the thickness of the sedimentary units, suggests that if they are a result of just one event then it would have been a flood much larger than any floods observed in the region since European settlement. There are many studies of slackwater deposits from around the globe published in the scientific literature. Examples of just a few of these include Heine (2004), Wohl (1992b, 1994), Kochel and Baker (1988) and Baker and Pickup (1987).

Plunge pool deposits

Plunge pools occur at the base of waterfalls. They are dish shaped features eroded by the impact of the water onto rock or sediments. Their size and depth are to a degree a function of the stream discharge and the height of the waterfall (Young, 1985). Streams draining plunge pools are usually graded to both the plunge pool and the trunk stream into which they flow. The height of the bed of the stream draining the plunge pool controls the average water level in the pool when discharge levels are sufficiently high. The volume of water in the pool will of course fall below the level of the bed of the draining stream when evaporation exceeds that of recharge. During high discharge events the water level in the plunge pool can rise suddenly causing overbank flows in the draining stream. Waves and currents are generated in the pool during high discharge events and these transport sediment, usually sands, from the floor of the pool to form high water stage beaches surrounding the pool.

Most plunge pools form at the heads of gorges where the waterfall, through knick point retreat, is in the process of extending a gorge into a highland. The streams draining the pool in these circumstances are confined within the gorge. In these geomorphic settings, high discharge events are responsible for depositing the beach surrounding, or to one side of, the pool; however, subsequent higher discharge events have the capacity to erode and remove that beach. The plunge pools examined by Nott and Price (1994, 1999) are different as they are not located at the heads of gorges. At Gunlom in Kakadu National Park and Wangi Falls in Litchfield National Park, Northern Territory, Australia (Fig. 3.4), the plunge pools are located at the base of long escarpments where the waterfall and associated stream have not incised a gorge into the highland. The streams, after plunging over the escarpments, turn sharply and follow the base of the escarpment. These streams drain to the side of the plunge pools. High discharge events cause a rapid rise in plunge pool water levels but the waters then spread out across the lowland plain rather than being confined within a gorge. The waves and currents generated in these plunge pools during floods also spread out some distance from the normal plunge pool beach and generate another beach, or beach ridge, that marks the stage level of these higher discharge events. Gunlom and Wangi Falls have two beach ridges at progressively higher elevations away from the normal pool beach. Luminescence (TL) dating of the sediments in the ridges at Gunlom showed that Ridge 1, closest to the plunge pool beach, was deposited between approximately 2000 years BP and present (Fig 3.7). This ridge still receives sediment during very high magnitude discharge events. Ridge 2 is approximately 3 m higher than Ridge 1, and 150 m away from this ridge and 300 m away from the plunge pool beach. Ridge 2 was deposited between 22 000 and 5000 years BP. Sedimentation on Ridge 2 was not continuous during this period but occurred during two distinct phases: the first for approximately 4000 years around the Last Glacial Maximum (22 000–18 000 years BP) and then again from approximately 9000 to 5000 years BP corresponding to the Holocene Climatic Optimum. Ridge 2 has not been accumulating substantial quantities of sediment, indeed any at all, since about 5000 years BP suggesting that floods have been significantly smaller in magnitude (5–7 times smaller) since this time.

The two ridges at Wangi Falls revealed a similar story to Gunlom (Fig. 3.7). Wangi Falls is approximately 300 km east of Gunlom so it would seem unlikely that the similarity in flood histories between the two sites is a coincidence or a function of a localised catchment phenomenon such as a temporary natural

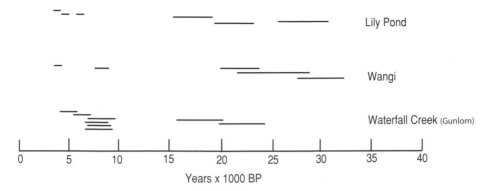

Years x 1000 BP

Figure 3.7. Time lines including uncertainty margins from thermoluminescence dates for plunge pool deposits at Lily Pond, Wangi and Waterfall Creek (Gunlom), Northern Territory, Australia. Note the close correspondence between dates during the Holocene Climatic Optimum (10–4 ka), the gap in sedimentation between 15 and 10 ka and the late Pleistocene sedimentation between 30 and 18 ka.

damming of streams. Ridge 1 at Wangi Falls returned luminescence ages extending from 3000 years BP till present and Ridge 2 produced ages from 30 000 till 4000 years BP. Like Gunlom, Ridge 2 at Wangi Falls shows a substantial hiatus in sedimentation between approximately 17 000 and 9000 years BP. Ridge 2 at Wangi Falls stands approximately 9 m above the present plunge pool beach and 150 m away. Major flood episodes also deposit sediment on Ridge 1.

A third site in the 'Top End' of the Northern Territory also revealed a near identical flood history to Gunlom and Wangi Falls (Fig. 3.7). Baker and Pickup (1987) identified a channel scabland on the sandstone plateau above Katherine Gorge. Inspection of the site showed that flood flows have exceeded the capacity of the 30 m deep section of Katherine Gorge in this reach and eroded a bedrock channel into the plateau surface. In one location this bedrock channel narrows significantly and immediately downstream of the constriction increased velocity flows have eroded a large plunge pool like feature (Nott and Price, 1999). A small (~3 m high) waterfall or step in the bedrock floor lies at the upstream end of the plunge pool and like the plunge pools at Gunlom and Wangi Falls a sand ridge has been deposited to the side of the pool. There is only one ridge at this location (Lily Pond) and no modern plunge pool beach. The ridge rises to 7 m above the plunge pool mean water level and the modern plunge pool beach probably forms the lower slope of the single existing ridge. Luminescence ages from the crest of the ridge and further downslope again showed a major period of sedimentation between 23 000 and 4000 years BP with a significant hiatus between 17 000 and 8000 years BP. Ages from the lower part of the ridge slope returned similar ages to those obtained from Ridge 1 at Gunlom and Wangi

Falls suggesting that the highest section of the ridge has not been experiencing sedimentation since approximately 4000 years BP.

All the plunge pool sites here in the Northern Territory highlight a synchronous pattern of flooding over the past 25 000 years. Nott and Price (1999) were able to demonstrate that these phases of enhanced flooding, especially that related to the HCO, were due to relatively minor shifts in average climatic conditions. The plunge pool records emphasise that the changes in regime and magnitude of flood flows may be disproportionate, with the latter (Holocene) experiencing much greater relative change than the former. Similar conclusions have also been observed in the USA by Ely *et al.* (1993) and Knox (1993). These flood records, though, were not in the form of plunge pool sediments.

Geobotanic indicators

Geobotanic indicators of palaeofloods can be grouped into four main classes: corrasion scars, adventitious sprouts, tree age and tree-ring anomalies (Wohl, 1995). Corrasion scars occur when part of the cambrium wood is destroyed by the impact of flood-carried debris. The wood tissue is scarred and this mark can provide a date for the flood event. The height of the scar can also be used to estimate the height of the flood waters although this may be a minimum stage height depending upon the nature of the debris impacting the tree. For example, if the debris were large rocks then it is unlikely that such rocks would have been transported near the water surface but at some depth below. Alternatively, if the impacting debris was something that normally floats such as tree branches then the corrasion scar may be an indicator of the flood stage height. It is important to assess the likely nature of the impacting debris before using this type of data to estimate the flood height.

Tree rings, like scar tissue, can also provide data on the age of a flood. Trees can be tilted by flood waters and this can cause eccentric ring patterns. Also, ring suppression and release patterns and variations in ring width can be caused by changes in river flow patterns. The age of trees growing on flood sedimentary sequences or in zones of erosion caused by high flood waters can also help date a flood. Likewise, the age of adventitious sprouts growing from broken tree stems due to floods can be determined by examining the ring patterns toward the base of the sprout.

Flow competence measures

The size of sedimentary clasts transported by flood flows can be used to gain a measure of the velocity, discharge and stream power of past floods.

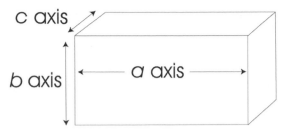

Figure 3.8. *a*, *b* and *c* axes of a boulder.

There is a proportional relationship between the size of the flood, in terms of discharge and stream power, and the size of clast able to be transported. Very large boulders can usually only be transported by very large floods and the characteristics of any such boulders left in a channel, as long as they have not fallen to their present position from slope processes, can be used to estimate the size of the past flood. The boulder characteristics that determine the magnitude of the past flood include the boulder size, particularly the length of the *a*, *b* and *c* axes (Fig. 3.8), the boulder rock density and boulder mass, the boulder shape and the position of the boulder prior to transport. This latter variable is difficult to determine but boulders in an imbricate position (where they are stacked against each other like shingle tiles), or where they are tightly packed in a group, require considerably greater forces to initiate motion than a boulder standing in an isolated position. Boulders are usually assumed to have been standing originally in an isolated position prior to transportation to their present position.

A particle will be entrained in the flow when the drag and lift forces acting on that particle exceed the force of resistance. Resistance is a function of the particle mass, shape and pre-transport position minus the buoyancy of the water covering a submerged particle. This can be expressed by the following equation

$$\text{drag force } (F_D) + \text{lift force } (F_L) > \text{resistance force } (F_R) - \text{buoyancy} \qquad (3.1)$$

Drag and lift forces are dependent on the difference in pressure in the fluid flow on opposite sides of the boulder (i.e. upstream and downstream sides). Drag force increases on the upstream side of a boulder as velocity increases. Likewise, as velocity increases there is a corresponding drop in pressure immediately above the boulder (Bernoulli effect) causing the boulder to be lifted upwards into the flow. Boulders are usually transported by rolling over their shortest axis. Boulders transported in a stream or in any fluid flow are usually, but not always, deposited with their long axis perpendicular to the direction of flow. In other words, the boulder is transported by rolling over its *c* axis and is deposited with its *a* axis perpendicular to the flow direction. So when a subsequent flow entrains the boulder the drag force will operate on the upstream face bounded by the *a* and *c* axes. This is an important assumption made in the equations used to calculate the flow velocity responsible for transporting a particle.

The drag force can be described as follows

$$F_D = C_D \cdot d^2 \cdot \left(\gamma_f \cdot V_b^2 / 2 \right) \tag{3.2}$$

Where C_D = drag coefficient, d = diameter of particle (usually the *b* axis), γ_f = specific weight of the fluid (g^{-1}) and V_b = flow velocity at bed.

The drag coefficient (C_D) is a function of the shape of the particle. Sphere shaped particles are easier to entrain than a cube. The smaller the face being acted upon by the drag force the higher the velocity required to initiate motion. Drag coefficients can be defined by a shape factor (SF) expressed as,

$$SF = d_S / \sqrt{d_I} \cdot d_L \tag{3.3}$$

Where d_S = diameter of *c* axis, d_I = diameter of *b* axis and d_L ≡ diameter of *a* axis.

The lift force can be expressed as follows,

$$F_L = C_L d^2 \left(\gamma_S \cdot V_b^2 / 2 \right) \tag{3.4}$$

where C_L = coefficient of lift and γ_S = the specific weight of the sediment. The lift force can only apply where the fluid flow covers the entire particle being entrained. Costa (1983) suggests that a lift coefficient of 0.178 be used as it represents the velocity one-third of the particle diameter above the stream bed or the point at which lift occurs.

The resisting force can be expressed as follows,

$$F_R = (\gamma_S - \gamma_f) g \mu \tag{3.5}$$

where μ = coefficient of static friction and g = gravitational constant. Static friction refers to the friction between the boulder and the channel bed. Costa (1983) notes that μ varies between 0.5 and 0.8 with a commonly accepted value of 0.7.

Simplifying equations (3.2), (3.4) and (3.5) and solving for bed velocity (V_b) gives,

$$V_b = \sqrt{[2 \, (\gamma_S - \gamma_f) d_I g \mu / \gamma_f (C_L + C_D)]} \tag{3.6}$$

The average flow velocity can be obtained by multiplying the bed velocity (V_b) by 1.2

$$V = 1.2 V_b \tag{3.7}$$

Williams and Costa (1988) and Costa (1983) outline the derivation of these equations in more detail and Wohl (1995) reviews some of the potential problems associated with using such equations. There are several potential limitations with using equations describing the relationship between stream velocity and boulder size to determine the discharge of a palaeoflood. Some of these include the variability of the velocity profile in streams and knowing which velocity is the most appropriate one for the movement of the clasts in question (stream velocity varies with flow depth in rivers). Also, the effects of particle shape and packing, boundary roughness and variable fluid density are not well understood. Often, however, boulder piles in streams may be the only evidence available from which to gauge the size of a palaeoflood. So in terms of hazard risk assessments, these equations can provide very useful information on the size of past events, especially when attempting to ascertain the largest relatively recent flood in a catchment.

Another way of expressing the relationship between stream flood parameters and boulder transport is by using stream power defined as

$$w = \gamma \, Q \, S/W = \tau V \text{(mean)} \tag{3.8}$$

where w = stream power per unit area of bed (W m^{-2}), γ is density of fluid, τ is boundary shear stress (N m^{-2}), V(mean) is mean flow velocity in m s^{-1}, Q = discharge (m^3 s^{-1}), S = energy gradient or water surface slope and W = water surface width in metres.

Boundary shear stress is determined by the duBoys equation

$$\tau = \gamma D S \tag{3.9}$$

where D = flow depth.

Using these equations, mean flow velocity can also be determined using

$$V \text{(mean)} = w/\tau = w/\gamma D S \tag{3.10}$$

Simplified relationships using these fundamentals have also been developed for stream power and clast transport by Williams (1983) as

$$w = 0.079 d^{1.27} \tag{3.11}$$

(where d is the length of the b axis) and for a threshold mean velocity by Costa (1983) which is the critical velocity at which boulder motion will commence. This can be expressed as

$$V_c \text{(mean)} = 0.18 \, d^{0.49} \tag{3.12}$$

Baker and Pickup (1987) used stream power relations to determine the flood flow frequencies required to move sandstone boulders as large as $4 \times 3 \times 1.5$ m^3 within the Katherine Gorge, Northern Territory, Australia (Fig. 3.4). They found that the 1% AEP (1 in 100 year) event with a discharge of 6000 m^3 s^{-1}, flow velocities of 7.5 m s^{-1} and flow depths of 15–45 m were able to transport these boulders and probably caused cavitation features such as pot holes to occur in the floor of the bedrock gorge. These flood flows, along with structural controls in the bedrock, were also responsible for the size and location of pool and riffle sequences in the Katherine Gorge. Pools and riffles are the zones of scour and intervening sediment deposition, or shallows and sometimes rapids, in rivers. Pool and riffle spacing is typically 5–7 times the channel width and in alluvial rivers, along with the planform of the stream, is usually seen to develop by the 2–3 year return interval flood. Bedrock channels, whose channel boundaries are considerably more resistant to erosion than alluvial channels, do not appear to conform to this relationship. In the case of the Katherine Gorge, the spacing of pool and riffle sequences appear to be determined, not by the 2–3 year flows, but by much more extreme flood events. Hence, pool and riffle dimensions within a bedrock stream system can also be used as a guide to the size of past floods. Wohl (1992a, b) found a similar relationship between extreme flood flows (palaeofloods) in the Burdekin and Herbert River Gorges in northeast Queensland and the location of pools and riffles, boulder sizes, erosional scour features (Fig. 3.9) and also slackwater sediment deposition zones.

Channel geometry

Like pool and riffle spacing, the wavelength of meanders in rivers can be used as a guide to the channel forming discharge. Meander wavelength, channel width and discharge all appear to be interrelated, such that as discharge increases so does channel width and wavelength. The width of an alluvial channel is approximately proportional to the square root of discharge (Q) and likewise meander wavelength also varies with $Q^{0.5}$. Which particular discharge is the most effective, i.e. mean annual flood, mean annual discharge, the 1% AEP discharge or most probable annual flood is a matter of slight controversy (Knighton, 1984). But there is little doubt that high channel forming discharges tend to produce longer meander wavelength (distance between crests or bends of the same orientation). The water carrying capacity, i.e. channel width and depth or cross-sectional area, of a channel also tends to increase as discharge increases. In alluvial rivers this can vary from flood to flood (i.e. a flood flow can erode the banks and bed of a channel in order to accommodate an increased flow capacity) but the average channel geometry is thought to be determined by the 2–3 year flood event. The same appears generally true of meander wavelengths in alluvial channels. Whether this relationship holds true for bedrock channels or, as has been shown for pools and riffles, tends to be related to much higher discharges,

Figure 3.9. Bedrock scour concavity in limestone, in a tributary (Broken River) of the Burdekin River, North Queensland. The scour concavity marks the height of past floods. Note the dog in the right foreground for scale.

remains to be thoroughly investigated. However, Dury (1973) argued that bedrock meander wavelengths are a function of more frequent flood events. Dury noted though that many bedrock streams have meander wavelengths that appear too large to have formed under modern high frequency flood discharge events. He suggested that this phenomenon is best explained by climatic change where the present climate is less humid and the size of floods has decreased to the present day. In other words, the large meander wavelength of the bedrock channels is due to a wetter climate in the past when the size of the 2–3 year flood was much greater. The modern streams, which display meander wavelengths smaller than that of the bedrock gorge in which they flow, were termed 'underfit streams' by Dury.

While the relationship between meander wavelength and climate change in bedrock channels is ambivalent, there is little doubt that high frequency floods

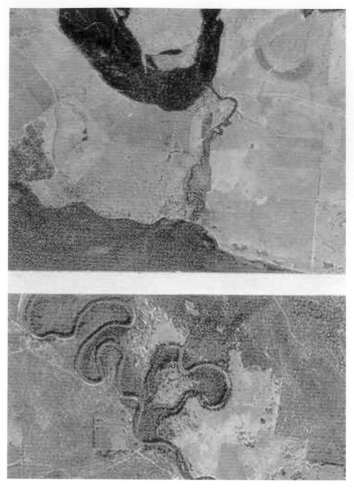

Figure 3.10. Aerial photographs of Green gully (Thule Lagoon Reach) (top) and modern Murray River (bottom) Australia at the same drainage area and scale. Green Gully was the ancestral Murray River prior to 12 000 years BP (Page *et al.*, 1996). Note the much larger meander wavelength of the ancestral Murray River (top image) compared to the modern Murray River. Photograph courtesy of Dr Ken Page, Charles Sturt University.

are responsible for meander wavelengths in alluvial channels. This is because alluvial channels are relatively easily eroded during a flood. In some locations, relict alluvial channels display meander wavelengths that are considerably larger than those forming in modern channels (Fig. 3.10). The suggestion is that these relict channels were formed by floods of greater discharge but with approximately the same frequency of occurrence as the floods forming the smaller channels today. The clear implication, therefore, is that flood discharges have decreased towards the present day. By dating sediments within the relict channels it is possible to constrain the age of this previously enhanced flooding regime.

Coral luminescence

Massive corals contain annual growth layers similar to rings in trees. Sometimes these growth layers display luminescent lines that can be seen when the sawn surface of massive corals is illuminated with long-wavelength ultraviolet (UV) light (Isdale, 1984; Lough *et al.*, 2002). Luminescent lines, previously known as fluorescent bands, were originally attributed to terrestrial humic substances incorporated into coral skeletons (Boto and Isdale, 1985). In this model, humic acids are incorporated into river flood plumes as sediments are eroded from river banks and floodplains during floods. Larger floods create plumes of water which extend well offshore and the humic acids carried by these plumes affect the outside layer of calcium carbonate or that year's growth layer on a massive coral. Hence, large floods will not only extend further offshore but also contain a higher quantity of humic acids and the resulting luminescence intensity reflects the size or discharge of the terrestrial flood. Using this technique, Isdale (1984) and Isdale *et al.* (1998) reconstructed river flood records for the Burdekin River, North Queensland for the last 350 years.

More recent research has cast doubt on the notion that the humic acids are responsible for the luminescent lines. Barnes and Taylor (2001) suggest that variations in the luminescence of the annual growth layer are more likely to be caused by variations in the growth of the coral skeletal architecture, which may in turn be influenced by the volume of freshwater entering the sea. The skeletal architecture is essentially a function of the size of the spaces between skeletal elements in the coral growth layer. Variations in this architecture probably arise from reduced calcification during periods of decreased salinity resulting from the dilution of seawater by freshwater run-off. This effect probably better explains situations where the occurrence of luminescence in corals that grow significantly offshore could not be influenced by terrestrial humic acids. Even if humic acids are not responsible, the skeletal architecture hypothesis still accommodates the idea that the luminescent lines record flood plumes into the ocean. Hence, the timing and intensity of the lines can be closely related to the timing and intensity of river run-off (Isdale 1984).

Lough *et al.* (2002) examined slices from 232 similar-sized colonies of massive *Porites* corals from 30 reefs on the Great Barrier Reef (GBR), Australia. The coral slices were viewed under UV light in a darkened room. For each colony and year, the appearance of the luminescent lines was graded into one of four categories:

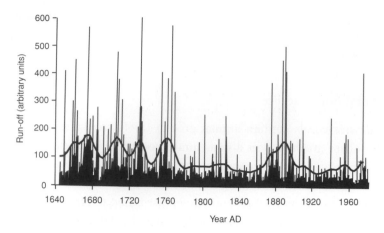

Figure 3.11. Coral luminescent line record of flood discharges from the Burdekin River, North Queensland, Australia (data from Isdale *et al.*, 1998).

0 = no visible line, 1 = faint luminescent line, 2 = moderate luminescent line and 3 = intense luminescent line. Reefs were divided into three groups: those that never recorded run-off, those that recorded run-off but not every year and those that recorded run-off every year. Lough *et al.* (2002) found that the average intensity of the luminescent lines was inversely related to the distance of the coral reef from shore and also to the average water depth between a reef and the mainland. They also found that annual variations in luminescence intensity were significantly correlated with instrumental measurements of annual river flow. This suggests that, on the Great Barrier Reef at least, luminescence lines are a reliable proxy of river discharge.

Figure 3.11 shows a 350 year record of luminescent lines from corals experiencing the flood plume from the Burdekin River, North Queensland. It is interesting to note that this annual record shows periods of relative quiescence in flood discharges between approximately AD 1770 and 1870, and the period from AD 1640 to 1770 was dominated by many more high discharge events than during any other century over the sampled time period. Except for the period around AD 1880 to 1910, flood discharges up to present have been relatively low for over two centuries. Such a record highlights the concept of hazard or event regimes and emphasises the need to check for non-stationarity when undertaking risk assessments. In this case, non-stationarity is clearly evident in this prehistoric record.

The largest known floods on Earth

High energy megafloods with discharges as great as $1 \times 10^7 \, \text{m}^3 \, \text{s}^{-1}$ occurred during the late Pleistocene when ice dammed lakes burst releasing huge volumes of water. Flood flows from ice dammed lakes are common in Iceland, but not to the same degree as those during the late Pleistocene. These types of floods are called jökulhlaups. The Pleistocene jökulhlaups occurred around the margins of the last glacial ice sheets in both North America and Asia. The evidence for these floods is in the form of channel scablands where large scale erosional features including scour hollows and cavitation features such as pot holes as well as bedrock scoured overflow channels occur. Also sedimentary sequences including 20 m diameter boulders and giant gravel ripple sequences have been documented for the channel scabland area of Washington State (Baker, 1973) and the Altai Mountains region of Siberia (Rudoy and Baker, 1993; Carling, 1996). In Washington, the largest glacial lake impounded about 2500 km^3 of water which was 600 m deep. After failure of the impounding ice dam, the outflow was so great that the flood transported and deposited graded sedimentary beds between 100 and 160 m high in a submarine rift valley 1000 km from the mouth of the Columbia River (Baker, 2002). These flood flows are similar in size and velocity to the world's major ocean currents such as the Atlantic Gulf Stream, the Kuroshio in the Pacific and the Agulhas current in the Indian Ocean. They are also similar in dimension to the megafloods that formed the outflow channels on Mars which were responsible for filling the Oceanus Borealis (the great ocean thought to have formed across the northern portion of Mars) and changing the atmosphere and climate of Mars from cold and dry to warm and humid several times over the last 4.6 billion years (Baker *et al.*, 1991).

Conclusion

Floods are one of the most devastating of all hazards globally. More than 75 million people are affected by floods each year, so considerable effort and expense is directed towards mitigating against this hazard. Despite their frequency of occurrence, many communities are still caught unaware by exceptionally large floods either because they have had no experience with events of this magnitude or did not realistically believe that such an event might affect them. Short historical records may give a false impression of the nature of the flood hazard for a region. Palaeohydrology can help in this regard by providing a valuable insight into the nature of the flood regime for regions as diverse as the humid tropics to deserts where river flows may be relatively rare. Palaeoflood reconstruction can be undertaken using a variety of techniques including slackwater sediment analysis, flow competence measures and stage height indicators such as scour lines, and various forms of debris. Coral luminescence banding has also provided an unusual and accurate reconstruction of floods in the Great

Barrier Reef region of Australia and here, like so many other records, it is apparent that large flood events do not occur randomly over time. Like many other types of hazards, floods tend to cluster into active and less active phases that may be decades to approximately a century in length. Recognition of these phases is possible only where records are long. Such records are an invaluable and, indeed, a necessary component of any flood risk assessment.

4

Tropical cyclones

Formation of tropical cyclones

Tropical cyclones form over tropical seas and oceans where the sea surface temperature is at least 27 °C. They are low pressure systems that develop a warm core or eye structure which is an area of subsiding air in the centre of the system. Around and towards the eye, air spirals inwards and upwards from the outer parts of the cyclone with the area of maximum uplift occurring adjacent to the eye. As the air converges inward, it is deflected to the left (clockwise) in the southern hemisphere and to the right (anticlockwise) in the northern hemisphere due to the Coriolis effect. The wind velocity increases towards the eye, approximately doubling as the distance from the eye is halved. This does not mean that larger diameter cyclones necessarily have stronger winds as the wind velocity is also dependent on the pressure gradient across the system with lower central pressures usually having a stronger pressure gradient.

Tropical cyclones can be likened to a thermodynamic heat engine where energy, due to evaporation from the ocean surface, is lost via thermal radiation after the air rises and diverges between 12 and 15 km altitude (Holland and McBride, 1997). If the air is allowed to continue to rise and diverge at the top of the troposphere (lower layer of the Earth's atmosphere) then air will continue to be drawn into the centre of the system and the tropical cyclone will intensify. There is a limit to this intensification. Energy in this system is also lost to the ocean surface by frictional coupling or by wind. The frictional losses by wind increase with the cube of wind speed, while energy gained from evaporation increases in a linear fashion with wind speed. Hence, as wind speed increases with the intensification of the tropical cyclone the energy loss increases relative to energy gain (Holland, 1997). Therefore, there is a maximum limit to which a tropical cyclone can intensify and this varies between ocean basins. In the Coral Sea region of the southwestern Pacific this figure is approximately 890 hPa central pressure and in the northwest Pacific it is more intense with maximum potential intensities of approximately 860–870 hPa theoretically achievable.

In addition to the uplift of air or convection over warm tropical seas, tropical cyclones also require atmospherically forced divergence of the uplifted air at high altitude. If this does not occur then convection of air will cease at the Earth's surface. The low-pressure system near the Earth's surface then requires a wind circulation system to develop. This often happens with a convergence of winds into the monsoon trough. These winds of course tend to be southeasterlies in the southern hemisphere and northeasterlies in the northern hemisphere. So if the monsoon trough lies in the northern hemisphere, the southeasterlies, after crossing the geographical equator, will tend, due to the Coriolis effect, to change direction into southwesterly winds. The reverse occurs with the northeasterlies when they cross the equator if the monsoon trough resides in the southern hemisphere during the southern summer. The convergence of this air into the trough can be the impetus for the development of a low level wind circulation. A tropical cyclone can then develop and intensify with continued uplift of air due to favourable conditions aloft. Shearing winds in the upper troposphere can be a major limiting factor to any further intensification of the tropical cyclone and indeed can cause dissipation of the system. These shearing winds are high-altitude winds that literally cut off the top of the cyclone causing a lack of 'outflow', or divergence of air in the upper troposphere, and the system shuts down.

Tropical cyclones typically form in tropical seas between 5° and 30° latitude. They usually do not form over the equator due to the lack of influence of the Coriolis effect, which is required to develop wind rotation around the system. Tropical cyclones frequently form on the western side of tropical ocean basins where warm waters typically accumulate because of the movement of ocean currents and equatorial easterly winds. Such conditions are enhanced when the Southern Oscillation is positive and at these times cyclones often occur closer to the coasts of the southeastern USA, Gulf of Mexico, Caribbean Islands and in the northwest Pacific near the Philippines and Taiwan. The same is true in the southwest Pacific when cyclones tend to form closer to the northeast coast of Australia. When the Southern Oscillation is negative, or an El Niño event occurs, tropical cyclone (hurricane) development in the southwest Atlantic Ocean is often reduced and the same is true in the far southwestern Pacific. However, at these times cyclone activity can increase in the central Pacific with the passage of the warm sea surface temperatures further eastward. Warm waters and cyclone development are not just restricted to the western sides of ocean basins, however, for the eastern Indian ocean off the northwestern Australian coast regularly experiences very intense tropical cyclones. This is due to the narrow, warm Leeuwin current that travels polewards along the western Australian coast from Indonesia.

The most intense tropical cyclones to form in historical times have occurred in the northwest Pacific Ocean near the Philippines and Guam. Here the strongest cyclone, Typhoon Tip, with a central pressure of 870 hPa formed in October 1979. Since 1958 four other cyclones have achieved central pressures below 880 hPa in this region. The Atlantic Basin has also had several very intense cyclones but not quite as intense as the northwest Pacific. Hurricane Gilbert in the Caribbean Sea had a central pressure of 888 hPa and the Labour Day storm of September 1935 in Florida, USA had a central pressure of 892 hPa. The South Pacific has historically had slightly less intense cyclones but Tropical Cyclone Zoe in December 2002 reached 890 hPa central pressure as it crossed the island of Tikopea at the south-eastern extremity of the Solomon Island group. Historically, no land crossing cyclones have achieved central pressures below 900 hPa in Australia but there were a number of cyclones crossing the Western Australian coast between 1998 and 2002 that had central pressures between 910 and 920 hPa. Even at these relatively modest, but still very intense (low) pressures, wind gusts around the eye wall can reach 300 km h^{-1}. Typhoon Tip had sustained surface winds of 340 km h^{-1} and the wind gusts would have been much stronger again.

There are some concerns that the rate of intensification of tropical cyclones may increase under a future altered climate (Emanuel, 1987). This is important because cyclones can intensify very rapidly and develop from relatively weak to very intense systems overnight, leaving communities relatively unaware of the impending hazard. The most rapid (measured) intensification occurred in 1983 when Typhoon Forrest in the northwest Pacific Ocean dropped its central pressure from 976 to 876 hPa in less than 24 h and estimated sustained surface wind speeds increased from approximately 110 to 280 km h^{-1}. Other tropical cyclones have been known to drop between 3 and 6 hPa central pressure per hour for periods between 6 and 24 hours.

Impacts of tropical cyclones

Wind, storm surge and waves are the major phenomena that cause damage to both natural and human communities during a tropical cyclone. Wind damage during a cyclone does not increase linearly with increasing wind speed. Maximum wind speed in a cyclone increases as the square root of the pressure drop and the magnitude of wind pressures is proportional to the square of wind velocity. So the damage that could be expected to occur to buildings will increase

Table 4.1 *Tropical cyclone intensity and damage costs*

Cyclone intensity Saffir–Simpson scale	Median damage (US$)	Potential damage (relative to category 1 event)
Tropical storm	<1 000 000	0
Category 1	33 000 000	1
Category 2	336 000 000	10
Category 3	1412 000 000	50
Category 4	8224 000 000	250
Category 5	5973 000 000	500

Table information from Pielke and Landsea (1998).

geometrically with increasing cyclone intensity. Pielke and Landsea (1998) suggest that a category 4 cyclone (on the Saffir–Simpson scale) may produce, on average, up to 250 times the damage of a category 1 cyclone. Table 4.1 from Pielke and Landsea (1998) shows the median damage caused by US hurricanes between 1925 and 1995. It is interesting to note that the median damage in US dollars increases from $33 million for a category 1 event to nearly $6 billion for a category 5 event.

Topography plays a major role in the ability of cyclonic winds to damage buildings. Buildings on ridge crests are more exposed than those on flat ground. The lee sides of hills can be prone to damage due to turbulence where wind velocities can increase by up to 20%. The design of buildings is also important in minimising damage. Generally, brick and concrete block buildings are more resistant to damage because they are less likely to be punctured by flying debris. The shape of roofs is also important; hip roofs, as opposed to flat or low-pitched roofs and those with gable ends, tend to experience less damage.

Storm surge is an elevation of the sea surface both ahead of and during the passage of a tropical cyclone towards the coast. The surge is really a long gravity wave with a wavelength similar to the diameter of the generating tropical cyclone. The surge wavelength is on average approximately 4 times the radius of maximum winds, which can be anywhere from 10 to 50 km; hence the surge wavelength can vary, depending upon the size of the radius of maximum winds, from 40 to 200 km. Storm surges, therefore, are not like the shorter wavelength or shorter period (periods of seconds) waves generated directly by the wind that sit on top of the surge. A surge can be likened to a raised, low angle, dome of water that inundates the coast. The water level associated with the surge at the coast can take several hours to reach its peak height, which usually coincides approximately with the time of cyclone landfall. The height of the surge is dependent upon a number of factors including the intensity or central pressure

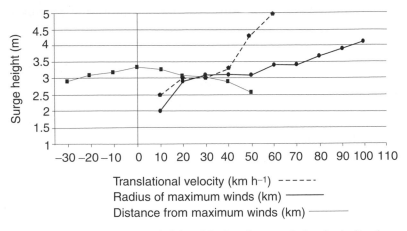

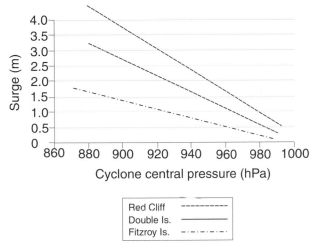

Figure 4.1. Variations in surge height with changing translational velocity, the radius of maximum winds and the distance from maximum winds in a cyclone for Red Cliff Point near Cairns, northeastern Australia (from Nott, 2003).

Figure 4.2. Variations in surge height between locations with different offshore bathymetries and coastal configurations (from Nott, 2003).

of the cyclone, the forward or translational speed of the cyclone, the radius of maximum winds, angle of approach of the cyclone track to the coast, the offshore bathymetry and the coastal configuration or shape of the coastline. Surge height increases as central pressure decreases or the cyclone intensifies. It also increases as the translational speed and radius of maximum winds increase, as the angle of approach becomes more perpendicular to the coast, where continental shelves are shallow and broad and also within certain shaped coastal embayments. Figures 4.1 and 4.2 show variations in surge height with translational velocity, radius of maximum winds and distance from the zone of maximum winds. Note that these changes are not uniform. For example, at Red Cliff Point in northeastern Australia surge height increases rapidly at translational velocities above 40 km h^{-1} and relatively slowly for velocities below this value. The same is true of the radius of maximum winds except that surge increases most rapidly when the radius of maximum winds is above 70 km. Surge height also varies with different offshore bathymetries and coastal configurations. Figure 4.2 shows three separate sites near Cairns in northeastern Australia. Two of the sites are islands (Double and Fitzroy Islands) and Red Cliff Point is on the mainland coast. The deeper waters surrounding islands restricts the rate of surge increase with decreasing central pressure, whereas broad shallow bays with a gently sloping offshore bathymetry can produce considerably larger surges with increasing cyclone intensity.

Storm surges by themselves can be very damaging to coastal buildings and infrastructure. When a surge inundates the floor of a building it can exert considerable stress on the structural integrity of that building. Given that 1 m^3 of water weighs 1 t, a floor area of 5 m^2 will experience 5 t weight when a surge reaches 1 m over the floor level. The weight of the surge alone can begin to result in the structural failure of a building when that surge exceeds one metre over the floor. Of course just the presence of water, even if it remains largely still or has very low velocity, can do substantial damage to household items just by soaking.

Storm surge is not the only component of inundation during a tropical cyclone. Total inundation of marine waters during a cyclone consists of the surge, tide, wave set-up, wave action and wave run-up. The surge is generated by the reduction in atmospheric pressure in the cyclone, by the onshore wind stress and the interaction of bottom currents with the sea floor. Atmospheric pressure unloading accounts for approximately 1 cm of sea-level rise for each 1 hPa reduction in air pressure. Hence, a pressure fall of 100 hPa to produce for example a 910 hPa central pressure cyclone will only result in 1 m of surge. But surges from cyclones of this intensity, depending upon the extent of each of the other surge producing factors, can result in surges in excess of 4 m. The onshore wind loading produces the majority of the surge in many instances. The onshore wind stress is, in a simplified description, the piling up of water against the coast due to the strong onshore winds. Water will only accumulate in this fashion where this onshore movement comes into contact with the ocean floor, or where the water is sufficiently shallow. The onshore wind stress component of the surge is minimal in deeper ocean waters such as those that may surround islands where submarine gradients are steep. A drop in ocean level, or a negative surge, can occur when the cyclonic winds are blowing offshore (from a mainland).

Storm tide refers to the level of inundation resulting from the storm surge and tide combined. This is not a simple addition of the tide height plus the surge height to give this level of inundation as the two do not interact or combine linearly because of the nature of bottom ocean currents. However, the tide together with the surge produce a greater level of inundation than the surge alone. Most eyewitness observations of marine inundation during a cyclone are observations of storm tides not surges, plus any wave set-up and possibly wave run-up. Wave set-up is the addition to the water column from broken waves and wave run-up is the uprush of those waves against an object or a sloping surface such as a beach. Wave set-up is normally regarded to be approximately 10% of the significant wave height (H_s) which is the average height of the highest one-third of waves. Wave run-up can vary considerably depending upon the gradient and composition of the slope surface that is overwashed. Wave run-up is an important component of marine inundation and is discussed in more detail later in this chapter.

Storm surges and storm tides can be measured directly by tide gauges if located sufficiently close to the zone of marine inundation. Eyewitness accounts can also be valuable as a guide to the level of inundation. However, people can often be under some stress during an intense cyclone and will usually be seeking shelter – hence eyewitness observations are not always reliable. Post-marine inundation surveys are valuable in those locations where tide gauges are absent to determine the height of the inundation and also to measure the impact upon the coast. Such surveys have been undertaken in Hawaii following Hurricane Iniki in 1992 (Fletcher *et al.*, 1995) and also along the Western Australian coast following several of the category 5 cyclones that occurred between 1998 and 2002 (Nott and Hubbert, 2005). These surveys involve topographically surveying the height of debris left by the inundation above the tide level at the time of the event at several locations along the coast. The results of one such survey following Tropical Cyclone Chris (915 hPa central pressure) in January 2002 can be seen in Table 4.2.

Nott and Hubbert's (2005) survey results highlight two very important aspects. First, erosion of the sandy coast including sand dunes occurred to the level of the maximum inundation level not just the level of the surge or storm tide. Second, it is possible to estimate, through numerical modelling of the surge heights during the event, the extent of wave run-up as noted in the difference in height column in Table 4.2. Both of these aspects are important for assessing risk to coastal communities from storm tides because many emergency strategies along with building guidelines only take into account the level of a surge or storm tide during a particular intensity cyclone. So for example the minimum building level above sea-level will be determined from the estimated height of a storm tide during various cyclone scenarios. Wave run-up is often not included in

Table 4.2 *Modelled and surveyed heights of marine inundation during TC Chris*

Location	Approximate distance from crossing (km)	Latitude (degrees south)	Longitude (degrees east)	Modelled surge and wave set-up (m)	Surveyed height of debris (m)	Difference in height (m)
Survey site 1	10	19.92	120.178	3.88	5.4	1.52
Survey site 2	20	19.9	120.236	3.18	4.1	0.92
Survey site 3	27.5	19.89	120.273	3.38	4.2	0.82
Predicted site of maximum surge	30	19.87	120.30	3.98	N/A	N/A

these plans, regulations or procedures. But wave run-up needs to be considered because the extent of erosion, and hence potential undermining of human built structures, will occur to the limit or elevation of that wave run-up, which can be considerably higher than the storm tide (see Figure 4.9).

Tropical cyclones also have a major impact upon the natural environment. Tropical forests, mangrove forests and coral reefs can all experience substantial mortality during a cyclone event; the extent of the damage often being dependent, but not always, on the intensity of the cyclone. The damage incurred by coral communities from category 5 cyclones can be devastating though it is unlikely that all corals are destroyed. Several studies have shown that there is substantial spatial variation in damage following cyclones (Connell, 1978; Connell *et al.*, 1997; Hughes and Connell, 1999) and different intensity storms can have varied effects. However, even when severe intensity cyclones do not cause extensive damage to coral communities they most probably weaken the substrate allowing severe damage to occur during subsequent less intense cyclones (Lirman and Fong, 1997). Hence, the frequency of the most intense cyclones is critical to the level of damage experienced by these communities over the longer term. The same is true for tropical rainforests especially where both the topography and canopy cover is uneven. Apart from human interference, it is unlikely that other disturbance mechanisms, although nonetheless extensive and significant (crown-of-thorns starfish, coral bleaching, phytophthora; Hughes, 1989; Hughes and Connell, 1999) would have the same level of impact as these extreme intensity cyclones. Given that these events occur at frequencies higher than the life span of many trees in the rainforest and corals on reefs (Nott and Hayne, 2001), it is reasonable to assume that the present character of these communities is a reflection of this disturbance regime. Tropical cyclones, therefore, may be instrumental in shaping the character of these natural communities and assist in promoting species diversity. Determining whether this is the case, however, requires much longer-term records than are often available from the historical record alone. This is also the case for assessing risk to human communities from these events. In such situations the prehistoric record can provide an invaluable insight into the trends and quasi-periodicities of tropical cyclone behaviour for a region.

PALAEOTEMPESTOLOGY AND THE PREHISTORIC RECORD

Palaeotempestology is the study of prehistoric tempests or storms. To date this has been confined to the study of prehistoric tropical cyclones but technically it could also apply to mid- and high-latitude events. Palaeotempestologists reconstruct past extreme events from sedimentary or erosional evidence left in the landscape as a result of storm surge and wave action. It is a relatively new branch of science which has, until recently, concentrated on the frequency of prehistoric tropical cyclones in tropical Australia (Chappell *et al.*, 1983; Chivas *et al.*, 1986; Hayne and Chappell, 2001, 2005), southern and eastern United States (Liu and Fearn, 1993, 2000; Collins *et al.*, 1999; Donnelly *et al.*, 2001a, b) and to a lesser extent throughout the South Pacific islands (McKee, 1959; Baines and McLean, 1976; McLean, 1993). In some situations, the magnitude of these past storms can also be estimated with reasonable accuracy (Nott and Hayne, 2001; Nott, 2003). By determining both frequency and magnitude it is possible to ascertain any quasi-periodicities in cyclone behaviour over the last 5000 years (Nott and Hayne, 2001). The historical record of cyclones is often, but not always (see Liu *et al.*, 2001 for 1000 year history in China), confined to the last 130 years or less. The longer-term insights offered through the study of palaeotempests allows more accurate risk assessments of the future impacts of these events to coastal communities.

The prehistoric record of tropical cyclones is largely restricted to the latter half of the Holocene, being approximately the last 5000–6000 years, or since termination of the Holocene marine transgression. While tropical cyclones would have formed during the period of lower sea-level between the present and last interglacials, any sediments deposited by these events are likely to have been reworked during the Holocene sea-level rise. Last interglacial storm deposits are probably also preserved in some locations, but to date no such deposits have been positively identified.

Sediments deposited during tropical cyclones take the form of ridges of coral rubble, pumice, shell or sand and shell, splays and sheets of sand incorporating marine microfossils, shells, coral fragments and lithic clasts, layers of sand within otherwise muddy or organic sediments in back-barrier lagoons and shell layers in fine-grained sediments in shallow marine environments. All of these

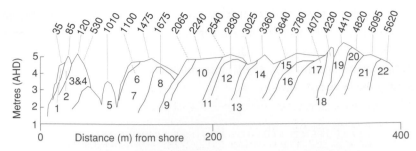

Figure 4.3. Stratigraphy and chronology for a coral shingle ridge sequence at Curacoa Island, Central Great Barrier Reef, Australia (from Nott and Hayne, 2001). Note the progressively increasing age of ridges with distance inland. Radiocarbon ages are mean (uncalibrated), marine reservoir corrected. See Nott and Hayne (2001) for uncertainty margins for each age.

sediments are deposited by surge and/or waves. Substantial quantities of sediment, mainly sand and occasionally isolated shells, can be transported inland and deposited by the high velocity winds of tropical cyclones. These deposits are difficult to recognise as they often occur in sedimentary environments where similar sand sized particles are deposited by normal aeolian processes.

Coral rubble/shingle ridges

Coral rubble ridges occur in locations where coral reefs occur close to shore. During the tempest, coral fragments are eroded from near-shore reefs by wave action and transported either onshore or offshore (Baines et al., 1974; Davies, 1983; Hughes, 1999; Rasser and Riegl, 2002). Fragments can also be transported from existing accumulations of coral rubble in the offshore zone. These offshore accumulations result from a number of erosional processes such as biodegradation and wave action during both storms and fair weather condition (Rasser and Riegl, 2002; Hughes, 1999). It is thought that the angle of the offshore reef slope plays a role in whether the eroded fragments are transported predominantly offshore or onshore. Steep reef fore-slopes favour offshore transport of fragments, often to depths of greater than 50 m which is too deep to be reworked and transported by storm waves. Shallow sloping, and particularly wide, reef fronts favour transport onshore and the formation of coral rubble ridges. However, some sites, such as Curacoa Island in the central Great Barrier Reef, Australia (Fig. 4.3) that are fronted by narrow, steep reef slopes have extensive coral rubble ridge development on land (Hayne and Chappell, 2001; Nott and Hayne, 2001). These sites with presumably minimal accumulation of coral rubble in the shallow waters of the reef and maximum accumulation of rubble in the deeper offshore waters below wave base (depth to which waves will entrain and transport sediment on ocean floor), suggest that the onshore ridges could have formed from predominantly live coral fragments broken off during the storm. At other sites, however, there can be little doubt that onshore ridges were formed from the reworking of existing accumulations of rubble in the shallow waters offshore.

It is difficult to know whether the rubble ridge is deposited gradually during the storm, or as one or a series of, sediment units moving landward from existing offshore accumulations. Scoffin (1993) has described coral rubble ridge building as ridges that 'have been transported and deposited like large asymmetric waves of sediment; material picked up on the seaward side is rolled up the ridge and dropped down the advancing slope' suggesting that an entire, or substantial part of an offshore accumulation of rubble is moved onshore as a single unit during the storm. If on the other hand ridges accumulate gradually, then the ridge could be assumed to increase in height over time during the storm. In this instance wave run-up may play a role in their formation.

It is likely that the height of an onshore coral rubble ridge is a function of the mean storm 'still' water level being the storm surge, tide and wave set-up combined, and possibly wave run-up. Ascertaining to what extent wave run-up is responsible for the height of the resulting ridge is important, as run-up can equal or exceed the height of the storm surge depending upon various conditions. Wave run-up is a function of significant wave height (H_s) and wave period/length, wave refraction/diffraction, bathymetry, beach slope angle and roughness and permeability of beach material (Neilsen and Hanslow, 1991). With very rough, coarse-grained, permeable substrates Losada and Gimenez-Curto (1981) note wave run-up can be 0.30–0.75 times that of run-up on a sandy, largely impermeable beach under the same storm conditions.

Observations of historical storm surge emplaced coral rubble ridges suggest that wave run-up may play an insignificant role. For example, a 3.5 m high ridge was deposited on Funafuti Atoll during Tropical Cyclone Bebe in 1971 (Fig. 4.4). The inundation accompanying the storm was 5 m above the level of the reef flat, or mean low tide level, and 1.5 m higher than the elevation of the resulting ridge crest (Maragos et al, 1973). A similar situation occurred at Mission Beach, south of Cairns, North Queensland, Australia, where an intense tropical cyclone struck in March 1918. Here a 4.5–5.1 m high (above Australian Height Datum or AHD) ridge of pumice was deposited by a surge as the cyclone crossed the coast. Eyewitness observations and results from numerical storm surge and wave models of the event, along with knowledge of the tide level at the time, shows

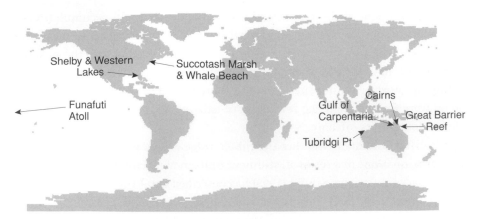

Figure 4.4. Location of palaeotempestology studies globally.

that the storm tide (surge plus tide) and wave set-up combined, amounted to an inundation level of 4.7–4.9 m (AHD). This suggests that wave run-up could have only contributed 0.2–0.4 m of the ridge height at it highest elevation. Elsewhere, where the ridge is only 4.5 m high (AHD), wave run-up does not appear to have contributed to formation of the pumice ridge.

It would appear, therefore, that in at least some instances the height of the crest of coral rubble and pumice ridges is close to or less than the height of the storm tide and wave set-up combined. This does not mean, however, that progressive accumulation or accretion of a ridge cannot occur during the storm. But it does mean that wave run-up may not always be an important process in ridge accretion. Determining to what extent wave run-up does play a role in ridge accretion is important when attempting to determine the intensity, or central pressure, of tropical cyclones responsible for the deposition of prehistoric ridges.

Coral rubble ridges contain a number of distinct sedimentary facies or units of sediment. These include storm beach face, berm, crest and washover facies (Hayne and Chappell, 2001). Beach face and berm facies include porous, clast supported, coarse biogenic shingle (rubble) deposits that occasionally dip sea-wards but are usually structureless. Crest facies are horizontally bedded and are finer grained than beach face deposits. Washover facies are bedded, dip landward up to 15° and sometimes contain imbricated clasts (imbrication is a sedimentary feature where particles are arranged in an overlapping shingle-like pattern dipping in one direction). Each of these facies or units combine to make a storm deposit. Storm deposits can be separated by 'ground surfaces' being lenses of pumice pebbles and a weak sooty or earthy palaeosol (ancient soil). These ground surfaces are really former ground surfaces or the surface of the feature that was exposed for sufficient time between individual cyclone events so that some soil development was able to take place. Ridges often contain only one storm deposit, but it is possible for two or more storm deposits to occur in one ridge. Careful excavation of a ridge is necessary in order to determine the number of storm deposits making up the ridge. Samples collected for geological dating from only one storm deposit, when more than one storm deposit is present, may bias the age determination of that ridge.

Coral rubble ridges often accumulate on the lee side of islands, presumably because on the exposed sides they are constantly removed by the largest or most intense tropical cyclones affecting a region. Leeward shores experience diminished wave energy, but the full effects of the surge. Because the wave energy is reduced, the likelihood of ridges being removed during subsequent cyclones is lessened. A number of ridges are sometimes able to accumulate over time where the preservation potential is particularly high. Curacoa Island on the Great Barrier Reef has 22 consecutive coral rubble ridges paralleling the shore on its north-western or lee side (Fig. 4.3). Individual ridges extend for over 100 m along shore and rise to over 5 m above the mid-tide level (the tidal range here is approximately 3 m). The ridges were deposited by successive cyclones so that new ridges are deposited seaward of the previously emplaced ridge. The age of the ridges increases progressively with distance inland. Curacoa Island is typical of many sites that preserve coral rubble ridges; however, not all sites retain as many ridges. Elsewhere, only one or two ridges paralleling the shore are preserved. This is often due to the site's exposure and the frequency of cyclones in the region.

The time interval between cyclones will determine the extent to which compaction and lithification or induration of the ridge can proceed. Over time, individual coral clasts within a ridge will begin to weather or break down, and in so doing provide carbonate that will progressively cement clasts together. Cementation usually begins within the core of the ridge, and it is common for loose fragments to remain on the crest and sides of the ridge several centuries after deposition. But, progressively over time, the ridge becomes resistant to further wave attack. However, looser fragments on the ridge crest can be removed and replaced by younger fragments if subsequent storm surges are sufficiently high to be able to overtop that ridge. In this fashion also, more recent storm deposits can be superimposed on older ones.

Prior to stabilisation, and depending upon the geomorphic setting, ridges can move inland by waves washing over the ridges and transporting clasts to the landward side of the ridge. The 19 km long, 3.5 m high and 35 m wide coral rubble ridge deposited by Tropical Cyclone Bebe on Funafuti Atoll continued to move inland and along shore for many years (Baines and McLean, 1976). Indeed, in some instances on coral atolls, the ridges will move inland across the reef flat and abut or lap onto existing ridges through normal, non-cyclonic, wave action. Individual storm or cyclone deposits and ridges will be more difficult to recognise at these locations compared to sites such as Curacoa Island where individual ridges often remain separated from each other.

Chenier and beach ridges

Cheniers are ridges composed predominantly, or entirely, of marine shell that has been deposited onto a mud substrate. They are also separated from each other by the mud substrate (Chappell and Grindrod, 1984). Beach ridges on the other hand can be composed of sand or sand and shell, and sometimes isolated coral fragments. Unlike cheniers, beach ridges are separated by sand swales, and ridge and swale topography together form a distinct and continuous sand unit which may be deposited onto any type of substrate. Cheniers and beach ridges are not restricted to tropical regions, and hence can form independently of tropical cyclones.

It is likely that most, if not all, cheniers are deposited by storm waves (or in some cases possibly tsunami), and if in the tropics these waves are likely to be due to tropical cyclones. Beach ridges on the other hand have been recognised to form by a number of processes, including deposition by swash during low or high wave energy conditions, or aggradation above the mean sea level by an offshore sand bar (Taylor and Stone, 1996) (Fig. 4.5). While all of these processes, and the beach ridges built by them, can occur independently of tropical cyclones, those ridges that occur well above mean sea level, and contain layers and/or beds of shell within tropical regions, are likely to have been deposited during cyclones.

Excellent examples of such ridges occur along the shores of the Gulf of Carpentaria, Australia (Rhodes et al., 1980). Up to 80 individual ridges, paralleling the shore, form a beach ridge plain here that extends inland in places for over 3 km. These ridges, along the eastern and southern shores of the Gulf, contain shell rich layers up to 1–2 m thick interspersed within medium- to coarse-grained sand (Rhodes, et al., 1980). The ridges rise up to 6 m above mean sea level (tidal range of approximately 2 m) and extend along shore for over 10 km in places.

A number of factors suggest that these ridges were deposited by storm surge and waves including: the height of these ridges above sea level, the presence of abundant shell layers within the ridge stratigraphy, that sea levels have not varied by more than 2 m in the region during the Holocene, and that the Gulf of Carpentaria is especially prone to the development of intense tropical cyclones because of its warm, shallow waters. Radiocarbon dating of the ridges by Rhodes et al. (1980) showed that the ridges increase in age progressively with distance inland.

Figure 4.5. Beach ridge sequence at Cowley Beach, south of Cairns in northeast Australia. Ridges and swales are differentiated by different coloured vegetation. Swales are noted by a lighter colour where Melaleuca trees dominate. These ridges are likely deposited by tropical cyclones. Photograph by Professor David Hopley, James Cook University.

Elsewhere, cyclone built ridges have existed only as ephemeral features. Coleman (1977) and Tanner (1995) describe cyclone built ridges near Florida, USA, which developed due to the accretion of an offshore bar and remained, until their eventual erosion by subsequent normal wave action, as isolated features separated from the mainland coast. Tanner (1995) argues that storm built ridges do not survive and hence are not preserved in the recent geological record. But the beach ridges from northeastern Australia suggest otherwise. Rhodes *et al.* (1980) suggest that the more inland ridges along the Gulf of Carpentaria probably date to the Pleistocene, suggesting that a pre-Holocene record of tropical cyclones may exist in this region. As yet, however, these ridges have not been dated because they no longer contain shells due to their weathering and decomposition over time.

Sand splays

Coastal sand dunes are often eroded, and diminished substantially in height, when surge and waves overtop them. The eroded sand can be transported inland as a splay or sheet that thins landwards. Similar deposits can also occur

Figure 4.6. Impact of Tropical Cyclone Vance (1999), Western Australia, upon coastal dunes. Here the impact was not as severe as in Fig. 4.7 as one row of dunes is still left intact except for scarping or horizontal erosion. Two rows of dunes were completely removed here. Dunes are 6–7 m high above beach level. Photograph by B. Hanstrum, Bureau of Meteorology, Western Australia.

when tsunami inundate sandy coasts (see Chapter 5). More studies of this style of tsunami deposit have been undertaken than the apparently similar storm deposits; as such the latter are not well documented in the literature.

An excellent example of a tropical cyclone deposited sand splay occurred when Tropical Cyclone (TC) Vance, with central pressure 910 hPa, crossed the Western Australian coast near Tubridgi Point in March 1999. Historically, this is the most powerful cyclone to cross the Australian coast. Winds around the eye were estimated at 300 km h^{-1} and Tropical Cyclone Vance generated the strongest ever recorded wind gust in Australia of 267 km h^{-1} at Learmonth, approximately 30 km from the eye. The zone of maximum winds struck a section of coast composed of a sand barrier comprising three rows of parallel dunes to approximately 6 m above the mid-tide level behind a wide sandy beach. In most locations along this coast, the first two rows of dunes were completely eroded and the sand transported away, presumably offshore, and the third or most inland dune row was eroded to form a steep scarp (Fig. 4.6). Precisely where the zone of maximum winds struck, however, all three rows of dunes were destroyed and the sand was transported inland as an extensive splay

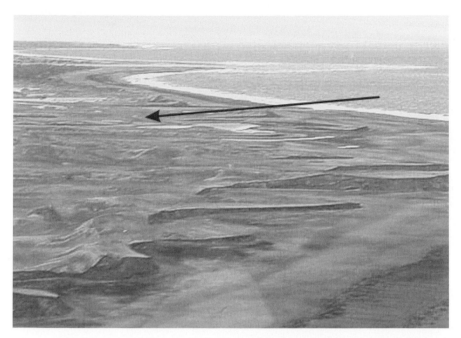

Figure 4.7. Area of greatest impact on dunes by Tropical Cyclone Vance. Here, three rows of 6–7 m high dunes were overtopped and completely removed by marine inundation. Note that vegetation which formerly consisted of grasses, shrubs and small trees was also removed. The large arrow shows the direction on onshore flow. Photograph by B. Hanstrum, Bureau of Meteorology, Western Australia.

approximately 400–500 m wide and 200–250 m inland from the first row of dunes (Fig. 4.7).

This sand splay decreased in thickness from 1.5 m immediately to the rear of the position of the former third row of dunes, to 0.75 m thick at its most inland extent. The splay terminated abruptly at a salt marsh where it was marked by a steep fronted (~30° angle) toe slope. Sediments within the splay were deposited as steep (~30°) tabular cross-beds (Fig. 4.8). Medium to coarse-grained sand occurred at the base of the unit, along with clasts of coral and shells, and graded upwards into medium- to fine-grained sand. The tabular cross-beds, along with other field evidence, suggested that the surge struck the coast with considerable force and moved inland as a reasonably high velocity bore. Such conclusions were supported by the presence of scour pits measuring up to 10 m long and 3 m wide on the lee (inland) side of trees, imbricated gravels and small boulders of lithic rock within the splay on its seaward side, and the shear volume of sand transported inland.

At present, too few studies have been undertaken to develop facies models that allow distinction between sand splays deposited by tropical cyclones and

Figure 4.8. Toe of a sand sheet deposited after denuding sand dunes during Tropical Cyclone Vance. Note the sedimentary structures (cross-bedded foresets, arrowed) in the foreground where the ridge has been excavated.

tsunami. The available literature suggests that tsunami sand splays are sometimes structureless, or have horizontal bedding, or have cross-beds indicating bi-directional flow (Goff *et al.*, 1998). The Tubridgi Point sand splay from TC Vance differs in having tabular cross-beds suggesting uni-directional flow (Fig. 4.8). Whether these different styles of sedimentation bear any specific relationship to tsunami versus storm wave processes, however, remains to be ascertained.

Washover deposits

The deposition of sand layers, up to tens of centimetres thick, in back barrier lagoons and swamps where fine-grained sediments are usually deposited, has been interpreted as evidence of storm washover events. Sediments within back-barrier lagoons are normally muddy, organic or fine grained. Interbedded sand layers within these fine-grained sediments can be due to storm surge and waves overtopping a sand-dune barrier, and transporting sand into an environment where it is not normally deposited. By geologically dating the sand layers a long-term history of cyclones in a region can be ascertained. It is important though to demonstrate that these sand layers are indeed from marine incursions during tropical cyclones and not from other sources such as rivers or indeed tsunami. Microfossils within the sand layers can help in this regard as different microfossil assemblages reflect their source location; tsunamis will tend to transport assemblages from deeper waters compared to tropical cyclone inundations.

Sand-layer stratigraphies from washover events due to tropical cyclones have been studied predominantly along the shores of the Gulf of Mexico and the southeastern and eastern United States. Liu and Fearn (1993, 2000) examined sand layers in lakes along the Florida and Alabama coasts. Donnelly *et al.* (2001a, b) studied similar deposits in New England and New Jersey along the US Atlantic coast (Fig. 4.4). All of these studies assume that the height and general nature of the barrier has remained unchanged over the length of the washover record. Such assumptions seem reasonable when separate sites some distance apart show the same chronology of events, or at least clusters of events, suggesting that some regional factor has influenced the behaviour of tropical cyclones at different times during the past. For example, in Liu and Fearn's studies of Shelby Lake, Alabama and Western Lake, Florida (Fig. 4.4), which are approximately 150 km apart, both records show a distinct clustering of apparently high-magnitude cyclones between approximately 3200–1000 ^{14}C years BP. These authors suggest that this clustering was due to a change in the latitudinal position of the jet stream over the mainland US, and as a consequence the location of the Bermuda High which moved further southwest. This caused Caribbean born tropical cyclones (hurricanes) to track farther westwards before taking their characteristic curve northward, resulting in more cyclones entering the Gulf of Mexico and striking US Gulf states. Before and after this time period, the jet stream was farther north and the Bermuda High farther northeast allowing tropical cyclones to strike the US Atlantic coast. Such assertions appear to be backed by a number of palaeoclimatic proxy records from various locations in the US.

Otvos (2002) was critical of the Liu and Fearn (2000) Shelby Lake washover record and he argued that most of the sand layers here were deposited, not from storm washover events, but from a variety of processes including redeposition from sand dunes that sit adjacent to and near the lake. Liu and Fearn (2002) in defence argued that the regional synchronicity of the washover record supports the cyclone surge origin of the sand layers and discounts the localised fluvial reworking of sand dunes.

While regional patterns in storm washover deposit records are probably the best indicator of a change in the nature of cyclone activity, and to a certain extent evidence that the deposits were due to washover events, it can never be assumed that the barrier being regularly overwashed has remained the same height over millennia. Hence, it is difficult to infer the magnitude of the cyclones responsible based upon the height of the present barrier unless a detailed chronostratigraphy of that barrier is undertaken. A chronostratigraphy is a quantitative assessment of the age or chronology of the layers, or various strata that comprise a geological unit. In this case the barrier sands, usually being sand dunes at the rear or landward side of the beach, will be composed of a number of units of sand that have accumulated, often due to aeolian processes, over time. If the dunes have remained well vegetated and largely undisturbed by wave attack there is no reason why such a barrier system should not remain at roughly the same height for some considerable period of time.

The dunes comprising the barrier at Western Lake, Florida form a continuous ridge, approximately 6 m high with some dunes rising to 9 m. For storm surge to overtop this barrier Liu and Fearn (2000) determined that the surge must have been generated by a high-magnitude cyclone (intense category 4 or 5 on the Saffir–Simpson scale). While there is a linear and proportional relationship between surge height and cyclone central pressure, it is possible for less intense cyclones to generate large surges when those cyclones travel at high forward speeds or have a large radius of maximum winds. Furthermore, the characteristics of the submarine topography and coastal configuration also strongly influence the height of a surge. These meteorological and geographical characteristics need to be considered in any study of palaeostorm deposits.

The long-term stability of overwashed barrier dunes remains one of the most uncertain aspects of this technique. Coastal sand dunes frequently erode when they are overtopped by surge and waves. The impact of TC Vance in Western Australia demonstrates clearly that several rows of high dunes can be completely removed during an intense cyclone. Examination of the impacts of other high-intensity tropical cyclones along the Western Australian coast between 1999 and 2002 (TC John, 915 hpa; TC Rosita, 930 hpa; TC Chris, 915 hPa) show that the extent and nature of dune erosion is a function of the height of the marine inundation (surge plus waves) relative to the height of a dune (Fig. 4.9). Dunes completely inundated by the surge and waves are likely to be removed, as occurred with TC Vance (Fig 4.9, stage 4). Dunes are reduced in vertical height, but not necessarily removed, when that dune is overtopped by wave run-up only, and not completely inundated. In this instance the waves wash over the dune, but the height of the 'still water' marine inundation (surge plus tide plus wave set-up) is less than the height of the dune (Fig 4.9, stage 3). Dunes experience recession, but not a reduction in vertical height, when the wave run-up does not reach the crest of the dune (Fig 4.9, stage 2). And dunes remain unmodified when the

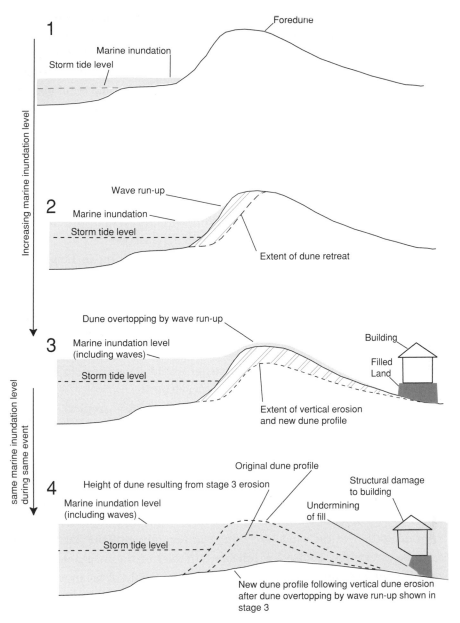

Figure 4.9. Schematic model of the relationship between dune erosion and the height of marine inundation. When inundation reaches the crest of a dune it will erode vertically. Where inundation is below the crest of a dune, it will erode horizontally.

marine inundation covers the beach only, and does not reach the dune, or only reaches the toe of the dune (Fig. 4.9, stage 1).

This relationship between dune height and the height of the marine inundation has also been recognised in the United States. The US Federal Emergency Management Agency (FEMA), in conjunction with the insurance industry, have developed a scheme to determine the extent of erosion likely along sandy coasts during hurricanes (FEMA, 2002). They estimate that when a frontal coastal dune has less than 540 ft^2 ($\sim$54 m^2) above the level of the 'still water' inundation (surge plus wave set-up but not waves), that dune will be completely removed. When the dune has greater than 540 ft^2 above the still water inundation that dune will experience erosional retreat, but no reduction in height. This formula is used to determine the risk associated with the construction of permanent buildings along a sandy coast by the US insurance industry.

The complete removal of dunes, and even the diminution in height of a dune barrier, will allow subsequent smaller surge and waves generated by lower intensity cyclones to penetrate inland and deposit sandy sediments into back-barrier environments. This situation may remain the case for many centuries and possibly longer. Eventually the dunes will rebuild to their former height, and over time will show no evidence of once being so extensively eroded. However, a chronological assessment of the dune barrier would show them to be much younger than some of the back-barrier washover deposits.

Accounting for extensive dune erosion during intense tropical cyclones remains one of the major caveats for the washover palaeostorm reconstruction technique. Any sand layers within back-barrier lagoon environments will be, to a certain extent, evidence of this erosion. It is likely that at least one of the sand layers in a palaeostorm sequence would have been deposited by an intense cyclone. But those stratigraphically above such a layer could have been deposited by less intense storms after erosion of the barrier dunes. The time between deposition of the sand layers and the rebuilding of the barrier dune needs to be ascertained before any reliable estimates of the magnitude of the storms responsible can be determined.

Long-term cyclone frequencies

Some of the earliest studies on the long-term frequency of tropical cyclones from prehistoric records occurred in the early 1980s through the examination of coral rubble ridge sequences along the Great Barrier Reef. Chappell et al. (1983) and Chivas et al. (1986) concluded that tropical cyclone frequencies have remained unchanged in the northeast Australian (Coral Sea) region over the last 3000–5000 years. A more detailed investigation of two of these sites (Curacoa Island and Princess Charlotte Bay) by Hayne and Chappell (2001) confirmed the conclusions of the earlier studies.

The frequency of cyclones throughout the late Holocene at these sites was determined by radiocarbon dating fragments of coral from the core of the coral rubble ridges. By dating most, if not all, of the ridges at a site the total number of ridge building cyclones over time can be ascertained. Of course the rate of ridge building is dependent upon the number of cyclones passing the area with time and also the rate of coral reef replenishment or regeneration. Hayne and Chappell (2001) concluded that the narrow, fringing coral reef, which was the source of rubble for the ridges on Curacoa Island, was able to regenerate after 80 years. This time period is considerably less than the apparent recurrence interval of ridge building (storm deposit) events of 280 years. The other ridge sites along the length of the Great Barrier Reef showed that ridge building events have occurred on average every 177–280 years (Nott and Hayne, 2001).

While the northeast Australian palaeorecord shows no change in the frequency of tropical cyclones at a multi-century scale, the US record shows a 2400 year period of increased cyclogenesis during the late Holocene. The record from Western Lake, Florida suggests that at least 12 high-intensity cyclones (category 4 or 5) struck this immediate section of coast over the past 3400 ^{14}C years, with only one event during the past 1000 years. The period between 5000 and 3400 ^{14}C years was also relatively quiescent. The record at Shelby Lake, Alabama shows five prominent sand layers were deposited during the past 3200 ^{14}C years and no sand layers were deposited between 4800 and 3200 ^{14}C years. Both records suggest a marked change in cyclone frequency over time with a distinct phase of heightened activity between approximately 3400 and 1000 ^{14}C years BP (Liu and Fearn, 2000, 2002).

Both techniques (coral rubble ridges and barrier washover deposits) of reconstructing long-term cyclone frequencies are limited in their ability to determine the magnitude of the cyclones responsible. In the case of the coral rubble ridges it was not known just which magnitude cyclones, or if all cyclones, affecting a region had remained unchanged in their frequency of occurrence. The same is also true for the barrier washover deposits because to date no assessments have been made of the chronostratigraphy of the overwashed barrier dunes. Hence, some of the sand layers may be from high-magnitude cyclones, others from less intense cyclones after the barrier dunes are eroded. However, if the barrier dunes have remained unchanged over the period of the palaeocyclone record, then the magnitude of the cyclones responsible can be bracketed broadly.

The intensity of prehistoric tropical cyclones

A method to determine the intensity of the prehistoric cyclones responsible for building ridges and eroding terraces in raised gravel beaches was
introduced by Nott and Hayne (2001) and Nott (2003). The height of these features is assumed to represent the minimum height of the storm inundation
during the event responsible. The elevation of these features is accurately surveyed to datum, and samples of coral and/or shell radiocarbon dated to determine the minimum height and times of inundation, respectively. The height
of this inundation is then related to the intensity of the palaeocyclone which
is determined through the use of numerical storm surge and shallow water
wave models. The models are used to determine the relationship between surge
height and central pressure for each location containing evidence of palaeo-
cyclones. Also, the relationship between the surge height and the translational
velocity of the cyclone, the radius of the maximum winds and the track angle of
the cyclone as it approaches and crosses the coast are determined. Model results
are compared to measured surge heights from recent or historical cyclone events
near the study sites. The central pressure of the cyclone responsible for formation of the ridge or terrace is determined by modelling the magnitude of the
surge plus wave set-up, and run-up and tide required to inundate the ridge or
terrace.

Nott and Hayne (2001) describe a series of lithic gravel terraces at Red Cliff
Point, north of Cairns in northeastern Australia. The terraces contain occasional
clasts of deposited coral and sit at a height of 6.1 m above Australian Height
Datum (AHD), or mean tide level (Figs. 4.10 and 4.11). The terraces have been
eroded by surge and wave action during tropical cyclones as they sit well above
the level of normal wave action and any slightly higher sea levels during the
mid-Holocene on a tectonically quiescent coast. Wave run-up across the gravel
beach at this site was measured during recent cyclones and wave set-up was
accepted as 10% of offshore significant wave height. By modelling the magnitude
of storm surge and offshore wave environment for a range of tropical cyclone
scenarios, Nott and Hayne (2001) calculated the central pressure of the cyclones
necessary to produce marine inundations (surge, plus tide, wave set-up and wave
run-up) to reach the level of each of these terraces. Nott and Hayne estimated
that the cyclone responsible for eroding the highest terrace would have had
a central pressure of approximately 900 hPa and generated a surge of 4 m,
offshore significant wave height of 5 m, wave set-up of 0.5 m and wave run-up
of 1.6 m. The tide height at the time of the prehistoric event is unknown but
can be estimated (at the 95% confidence level) to have occurred within the 2σ
probability tidal range of the frequency distribution nodal tide curve for each

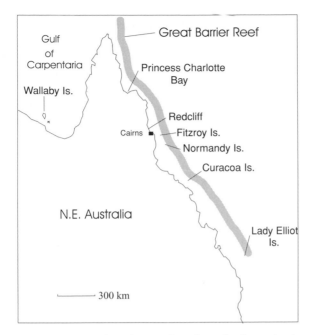

Figure 4.10. Location of study sites along the Queensland coast, Australia.

site. This tidal range effectively forms the uncertainty margins associated with the mean central pressure of the cyclone.

By assessing the magnitude of prehistoric tropical cyclones a more realistic view of the long-term frequency of these events could be ascertained. Nott and Hayne (2001) found that the coral rubble ridge sequences along the Great Barrier Reef were primarily recording only the largest or most intense cyclones, and were not registering the smaller less intense events over time. This stands to reason because any ridges that were built by lower magnitude events would have been removed by the more extreme events. This phenomenon is analogous to a dirty stain around the perimeter of a bathtub, for only the highest bath waters will leave a stain or ring and the shallower bath waters will be removed by the higher or deeper bath waters. Conversely, if shallow waters occur in the bath for some time after the last deep-water bath, the subsequent shallow waters will not remove the high water stain and shallow water stains will also be preserved. Hence, with coral rubble ridges, the highest ridges have the greatest preservation potential. The same is true of the raised gravel beach terraces at Red Cliff Point except here the lower elevation terraces have been preserved recording subsequent less intense cyclones.

These records, for the main part, tend to only preserve one spectrum of the cyclone climatology (i.e. the most extreme events). Nevertheless, these records are useful for they show how often the most extreme events occur. These events

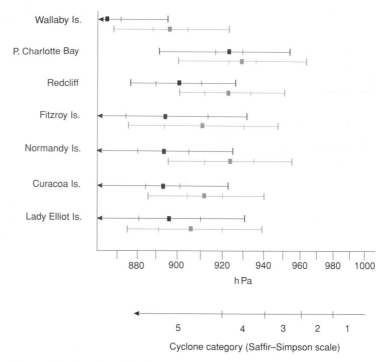

Figure 4.11. Intensity of cyclones responsible for depositing coral shingle ridges at locations shown in Fig. 4.10. Note that the intensity range corresponds to a 95% uncertainty margin and is determined by the tidal range over a full nodal tidal cycle, i.e. the most intense values correspond to a cyclone striking at the lowest astronomical tide, LAT (extreme values at a lower margin are unrealistic because they likely exceed the central pressures possible, hence the cyclone would not have occurred at LAT). (From Nott and Hayne, 2001.)

along the Great Barrier Reef were very intense category 4 and category 5 cyclones (Fig. 4.10). The prehistoric record shows that at virtually all sites along the Great Barrier Reef (seven in total) (Fig. 4.10), the average return interval of these intense cyclones is between 200 and 300 years over the past 5000 years (Nott and Hayne, 2001).

The possible caveats to the technique for determining the magnitude of prehistoric tropical cyclones, are the lack of inclusion of the influence of hydroisostasy on the present-day elevation of the prehistoric cyclone deposits, and the imprecision in calculating the wave run-up during the prehistoric event. Nott and Hayne (2001) and Nott (2003) recognise these possible caveats, and note that sea levels have fallen, due to hydroisostasy, over the period of the prehistoric record. However, if this is taken into account in the analysis of the intensity of the prehistoric storms, the older deposits (coral rubble ridges deposited between

3000 and 5500 years BP) should be higher in elevation than the younger more shoreward ridges. Topographic surveying of the ridges shows that this is not the case. The coral rubble ridges compact over time, and there is evidence also to suggest that the ridge crests have been overtopped by storm surge and waves, further reducing their height since deposition. This reduction in elevation of ridge crests is accepted by Nott and Hayne (2001) as more than compensating for the effects of sea-level fall since termination of the Holocene transgression. Also, if sea-level fall was taken into account it would suggest that the ridges deposited since approximately 2500–3000 years BP, or since sea levels stabilised at their present height, would have been deposited by more intense cyclones. Such an interpretation would suggest that these storms were becoming more intense with time.

Wave run-up is highly variable, depending upon a range of factors during an intense cyclone. Nott and Hayne (2001) undertook direct measurements of wave run-up onto the gravel terraces at Red Cliff Point, and over the coral rubble ridges and beaches along the Great Barrier Reef, during cyclones to determine a relationship between run-up and offshore significant wave height. These measurements were not made during cyclones as intense as those calculated to have occurred during prehistoric times, so an assumption of linearity between less and more intense storms was made. In order to compensate for this, Nott and Hayne (2001) undertook their modelling using extreme values for some of the cyclone parameters such as the radius of maximum winds R_m and the forward speed of the cyclones. Using such extreme parameters meant that they calculated central pressures for the prehistoric cyclones that were probably much higher (weaker) than those that occurred in reality. For example, a cyclone with a radius of R_m of 30 km and forward speed of 30 km h^{-1} will need to be much more intense than one with R_m of 60 km and forward speed of 40 km h^{-1} to produce a given height surge (Nott, 2003). By choosing the latter group, instead of the more likely former group of parameters, means that in order to deposit a coral rubble ridge of a certain height the modelled cyclone will have a higher (weaker) central pressure than that most likely to have occurred. In this manner, uncertainties in the extent of wave run-up can be compensated for in the final calculations of cyclone intensities. However, the only real solution to this problem is to obtain direct wave run-up values for extreme intensity cyclones, and this research is presently underway.

By recognising that prehistoric cyclones were more intense and frequent than that suggested by the historical record alone, raises the key question of whether these extreme intensity cyclones are occurring randomly with time, i.e. does this record suggest that on average the cyclones occur every 200–300 years. Or, are there centuries or several decades within this period when cyclones are

more active and likely to occur, and similar periods when they are less likely
to occur? It is difficult to answer these questions definitively at this stage but
some research suggests that there may be periods when cyclogenesis is active
and at other times not. The 1900s in Queensland, Australia appear to have been
relatively quiescent compared to the early 1800s when many more very intense
cyclones struck the coast (Nott, 2003). Quasi-periodicities also appear to be evident in the US historical hurricane record. The frequency of all hurricanes in
the tropical north Atlantic region did not vary over the last century but the frequency of intense storms (Saffir–Simpson categories 3–5) was variable and cyclic
at decadal scales (Landsea, 2000). The 1000 year record from southern China also
shows decadal episodes of enhanced cyclogenesis with two periods between AD
1660–1680 and AD 1850–1880 experiencing a much higher frequency of cyclones
occurring than at other times (Liu *et al.*, 2001).

Risk assessment of tropical cyclones using historical versus prehistorical records

There is little doubt that prehistoric data do not record the magnitude
or frequency of cyclone events as accurately as modern instrumented data. However, by its very nature prehistoric data often preferentially record the most
extreme events, as subsequent lesser magnitude events are unable to remove
this information from the landscape. Hence, when extreme magnitude cyclones
have occurred in the past they can often be clearly differentiated from those in
the historical record, for the palaeocyclone surge deposits and erosional terraces
sit much higher above sea level, or farther inland, than the highest recorded historical event. Likewise, while prehistoric chronologies derived using geological
dating techniques are less precise than historic chronologies, the uncertainty
margins of the technique (radiocarbon dating) allows a maximum possible age
(at the 95% probability level) to be determined. The beginning of the historical
record provides a minimum age if no events of these magnitudes are recorded
historically. Hence, while the prehistoric data have obvious shortcomings there
is no doubt that it is precise enough to reasonably determine when these events
occurred, and the extent to which they differ in magnitude to those recorded
historically. When prehistoric data register events larger than that seen in the
historic record it becomes a very valuable source of information because it shows
that the historical record does not include the entire range of cyclone events
likely in a given region. This has substantial implications for risk assessments.

Few comparisons have been made between the historic and prehistoric records
of cyclones for a region. Murnane *et al.* (2000) compared model predictions of
the probabilities of occurrence of wind speeds during tropical cyclones, based on
the historical record, with the palaeocyclone record from the washover deposits

in Florida (Shelby and Western Lakes), New England (Succotash Marsh) and New Jersey (Whale Beach) (Fig. 4.4). Good agreement was found between both types of records when the return probabilities of the prehistoric record were averaged in the comparison. There were, however, distinct periods of time during the late Holocene in these regions of the US when the annual landfalling probabilities of tropical cyclones differed substantially from the average. For example, the probability of an intense cyclone occurring per year based upon the average of the 5000 year record from Western Lake, Florida was approximately 0.35%. Whereas the period between 3400 and 1000 ^{14}C years BP at this site had an annual probability of cyclone occurrence of 0.5%, and the probability for the last 1000 years was much lower at 0.1% (Liu and Fearn, 2000). Interestingly also no cyclones of the intensity registered in the prehistoric record have struck the Western Lake region during the last 130 years. So which annual probability is the more realistic for risk assessment purposes?

Long-term records often show non-stationarity where periods or regimes exist that differ from each other in terms of event magnitude and frequency. Long-term records, therefore, remind us that the assumption that the relationship between the magnitude and frequency of tropical cyclones remains constant over time is not always true. Examination of cyclone palaeorecords allows us to identify the nature of regime changes and, therefore, make more accurate predictions of future risk. All too often, however, the prehistoric record, and hence recognition of hazard regimes, is overlooked in risk assessments of tropical cyclones. The northeast Australian region is a case in point.

When observational bias is removed, the record of tropical cyclones in the Australian region shows no appreciable change in variability over the last 130 years. The longest complete data set of land falling cyclones occur from northeast Australia, and here the usual pattern of many more lower magnitude events occurring compared to higher magnitude ones exists. The record shows that during this time period only one category 5 and two category 4 cyclones struck the east coast of Queensland, and each of these occurred prior to AD 1920.

The rarity of these extreme events since this time has led to a relatively blasé attitude towards the severity and consequences of such a hazard in some regional centres. Indeed, it could be argued that the lack of familiarity with such extreme events has led to the assumption that they are unlikely to ever occur within the lifetime of many individuals. This has led to the development of inadequate policies governing the location of buildings relative to sea level. Now there are numerous buildings less than a few metres above mean sea level, and well within the zone of possible storm tide inundation.

Using the historical data set, several studies have analysed the recurrence intervals of cyclones in this region, and through extrapolation estimated the frequency of the most extreme events (Harper, 1998; McInnes et al., 1999). There is little doubt about the veracity of the statistical methods and numerical models used in these types of analyses. However, when the observed cyclone frequency and characteristics are extended over several centuries by incorporating the prehistoric record of these events, there is a considerable difference in the estimated return intervals of severe magnitude cyclones (storm tide inundation) for the same region (Nott and Hayne, 2001; Nott, 2003).

Analysis of the prehistoric record near Cairns, Queensland show that during the period AD 1800–1870, the latter being the date of first European settlement, three cyclone events occurred producing storm tides between 2.52 and 4.51 m AHD. The highest historic storm tide was 2.5 m. Because there have been no inundations of this magnitude (>2.5 m) recorded or observed in this region since AD 1870, the prehistoric evidence suggests that the incidence of severe cyclones during the 19th Century was much higher than the 20th Century. When these data are combined with the historic record of inundations, the magnitude of the 100 year return interval storm tide (including wave set-up) increases by approximately 1 m (Nott, 2003).

The last few centuries have seen a regime shift in the occurrence of tropical cyclones crossing the coast in north Queensland. The causes behind this regime shift are not yet known, but they are likely to be climatic. Even in the absence of an understanding of the cause of the regime shift, it is still important for this shift to be considered in risk assessments. Yet risk assessments of tropical cyclone impacts to date in this region have been entirely based upon the historic record, and hence the present regime (Granger *et al.*, 1999). The implications of this approach for assessing building exposure to storm tide, amongst other important factors, are potentially serious. Based upon the variability of cyclones within the present regime, and at 1999 building patterns and levels, approximately 4% of housing and accommodation buildings would be inundated by storm tide during the 1 in 100 year event. Under the same scenario 17% of business and industry buildings would be inundated. If a return to the previous regime of cyclone variability were to occur, however, the percentage number of housing and accommodation buildings and business and industry buildings inundated by the 1 in 100 year event increases by greater than six times and four times, respectively. Given that these cyclone regimes appear to be century scale in length, it is entirely possible that a shift in regime may occur in the near future.

Future developments in palaeotempestology

Sediment deposited by wind during tropical cyclones is difficult to differentiate from normal aeolian deposits, or those deposited by winds generated during less intense meteorological conditions. Forests, however, can be impacted substantially during a cyclone through defoliation and tree fall (Boose *et al.*, 1994). As a consequence, early stage successional plant species will grow in place of those plants destroyed during the intense cyclone, and new pollen types will be distributed throughout an area. Pollen in sediments, therefore, may record episodes of forest destruction and the introduction of different species. Some of the major challenges facing future research in this area will be to differentiate the impacts of fire and human activities on forest decline from the impacts of tropical cyclones. Furthermore, the accuracy of such pollen records is also dependent upon the frequency of cyclone events of sufficient intensity to cause substantial damage to a forest, and allow pioneer species to grow instead. Otherwise, several events may occur before a forest has time to develop to a mature state, and the change in the pollen record within a sedimentary sequence will show only one event (Malmquist, 1997).

Variations in isotope ratios within speleothems (deposits of calcium carbonate within caves) also hold promise as a new technique to identify the passage of prehistoric tropical cyclones. Tropical cyclone rain is isotopically lighter than rain produced by other types of storms. This is because rain within tropical cyclones experiences little re-evaporation during its descent, whereas considerable re-evaporation occurs during normal convective showers. As a result, isotopic fractionation of hydrogen and deuterium occurs in normal showers, and little in cyclone generated rain (Ehhalt and Östlund, 1970). On the other hand considerable fractionation of oxygen isotopes occurs in tropical cyclones (Lawrence and Gedzelman, 1996). Changes in these isotopic ratios can be recorded over time, providing no contamination of the rainwater occurs between it falling on the ground and its incorporation into speleothems. Similar, isotopic ratios recording tropical cyclones may also be discernible in tree rings (White *et al.*, 1994).

Beds of sand and shell within otherwise finer-grained sediments on continental shelves may also record tropical cyclones. However, as with terrestrial deposits, it is difficult to distinguish between those sediments deposited by tsunami and storm waves. Keen and Slingerland (1993) compared model results of simulated tropical cyclone winds, currents, waves, and sedimentation patterns to observations of the actual storms. They found that the average thickness and lateral extent of the simulated storm sediment beds increased non-linearly with increasing wind speed. The configuration of the coast, however, caused considerable variation in the thickness of the beds between locations on the continental shelf. So far, this technique is useful for determining the frequency of tropical cyclones over an area, but the relationship to cyclone intensity remains less certain at this stage.

Conclusion

Palaeotempestology, the study of prehistoric storms, offers much greater insight into the nature of storms likely to impact a region than short historical records alone. Patterns and periodicities of prehistoric storms can be reconstructed from sediments, both organic and lithic, deposited by surge and waves at elevations or distances inland beyond the reach of normal marine processes. Erosional terraces can also be left in the landscape by the same processes. Techniques such as isotopic variations in speleothems, tree rings and variations in pollen species within sediments also hold promise as tools in deciphering long-term records of prehistoric tropical cyclones.

Millennial and century scale periodicities have been identified from sedimentary records in the south and southeastern USA and northern Australia, respectively, and such patterns highlight regime changes in the magnitude–frequency relationship of tropical cyclones over time in these regions. Recognition of these regime changes can really only be achieved through the use of the prehistoric record of these events where historical records are relatively short. Identifying the length of regimes, and mechanisms causing regime shifts, will become increasingly important for making more accurate predictions of the levels of risk and exposure as human populations, urbanisation and tourism grow rapidly along tropical and subtropical coasts.

5

Tsunamis

Tsunami characteristics and formation

Tsunamis of substantial size are waves of tremendous power and have brought great devastation to many coastal human populations around the globe. They have the ability to surge onto dry land beyond that reached by normal (wind generated) ocean waves. Because they are rare events, coastal populations have often settled in places that are occasionally, and surprisingly for these people, inundated by these waves. They have consequently acquired a reputation for the damage they can do to coastal infrastructure and the deaths to many who are swept away.

Tsunamis are very different from wind generated ocean waves. Tsunamis are long-period, shallow-water waves. They have periods (time between successive peaks or troughs) of many minutes compared to wind- and storm-generated waves which have periods of seconds. Because of their long period, the distance between tsunami wave crests can be many kilometres unlike the tens of metres for wind-generated waves. This results in the tsunami effectively touching bottom or the sea floor across even the deepest ocean basins. This is why tsunamis are referred to as shallow water waves. Shallow water waves occur where the water depth is less than half of the wavelength; the period and wavelength of a wave in the ocean is related to the velocity or celerity of the wave. The equation describing the celerity of shallow water waves in the ocean is

$$C = (gd)^{0.5} \qquad\qquad (5.1)$$

where C = celerity, g = gravitational constant (9.8 m s^{-2}) and d = water depth.

This equation shows that the deeper the water, the faster the wave will travel. As storm waves have much smaller periods, their wavelengths are much shorter than the water depth until they reach the shallow waters of continental shelves or reefs surrounding oceanic islands or atolls. Hence, they travel as deep-water waves for much of their existence and travel at relatively low velocities across oceans. Tsunamis travel much faster than storm- or wind-generated waves and can attain velocities of up to 800 km h^{-1} in the deepest ocean basins. Unlike storm waves, tsunamis often don't break until they reach shore. Hence, their greater velocity and period, and unbroken state until shore means that tsunamis are able to penetrate further inland. This is why they are able to wreak considerably greater damage to human communities and structures on the coast, and also the coastal landscape, than storm waves.

Tsunamis can be generated by a number of mechanisms including:

- submarine earthquakes;
- landslides into and below the ocean surface;
- volcanic eruptions;
- meteorite/asteroid impact into the ocean; and
- atmospheric disturbances either generated by volcanic eruptions or weather related phenomena such as squall lines and tropical cyclones.

Each of these mechanisms can produce different tsunami characteristics. Earthquakes generate tsunami through the rapid displacement of the ocean floor and hence the water column above. Earthquakes typically occur due to movements along faults, but not all fault movements and earthquakes produce tsunami. Faults vary in their structural characteristics. Faults that generate a vertical displacement of the ocean floor and water column such as dip/slip faults and thrust faults produce larger tsunamis than strike faults. Most larger and destructive tsunamis are generated by earthquakes greater than 7 and particularly 7.5 on the Richter scale. Often the magnitude of the earthquake, particularly if it occurs at shallow depth (<100 km), determines the size of the tsunami. Certain types of earthquakes appear more effective at producing tsunamis. Those that do not display a large peak energy release, but rather release their energy slowly due to the slow rupturing of the fault, can cause the formation of very large tsunamis. Such earthquakes are called tsunami earthquakes and can often occur without many people noticing them, or at least not becoming too alarmed because of their relatively low intensity. Earthquake-generated tsunamis are often composed of a number of individual waves that can be separated by distances of several hundred kilometres. It is not uncommon for more than five waves, and sometimes ten or more waves, to occur during the one tsunami event (Fig. 5.1).

Not all tsunamis are the same size, and the largest wave can occur in any position in the wave train. Like most tsunamis, earthquake-generated tsunamis are often less than 1 m high in the open ocean, but the wave height increases substantially as it travels across the continental shelf and slows in velocity. The

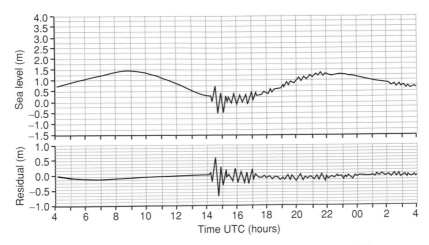

Figure 5.1. Example of a tsunami trace from a tide gauge record, Port Villa, Vanuatu, November 1999. Note the number of waves in the tsunami wave train.

steep topographic gradient of the shore relative to the wavelength of a tsunami allows the wave to surge onto the land at tremendous speed. For example, the Aitape tsunami in Papua New Guinea on the 17th July 1998 travelled across the sand barrier at Sissano Lagoon at approximately 80 km h^{-1} (the wave was 15 m high as it travelled across the beach). The destructive capabilities of such a high-velocity wave across dry land are highlighted when it is remembered that moving water in such situations has 1000 times the force of air travelling at the same velocity. Though strictly not an earthquake-generated tsunami, as it appears to have been generated by a submarine landslide which itself was caused by an earthquake, the size and velocity of the waves during this event were catastrophic to the local people living on the sand barrier (the official death toll was 2200 people but it was quite possibly much higher) (Kawata *et al.*, 1999).

Tsunamis have also travelled vast distances across oceans to impact shores far removed from the generating earthquake. The most famous of these 'telet-sunamis' is the Boxing Day (26th December) tsunami of 2004 that impacted a number of countries in the south Asia region and also east Africa (Somalia and Kenya). A magnitude 9.3 earthquake occurred just to the west of Banda Aceh off the northern tip of Sumatra. Within minutes tsunamis 10–15 m high impacted the west coast of northern Sumatra resulting in the deaths of over 100 000 people. The tsunami wave train radiated northwards into the Bay of Bengal, south to Western Australia and Antarctica, west towards India, Sri Lanka, the Maldives and East Africa and east towards Thailand and Malaysia. The eastward travel-ling waves first approached the coast of Thailand as a trough causing the water to initially recede from the shore followed minutes later by the first wave crest. Waves travelling west first impacted the coast of Sri Lanka, India and the Maldives as a crest followed by the trough. There were at least six waves of substantial size that impacted the coasts. In the state of Tamil Nadu in southeast India the largest waves were between 4 and 6 m high as they crossed the shore. Similar wave heights were experienced in Sri Lanka and parts of Thailand. Many who died were from Europe and elsewhere around the globe as they were holidaying in popular coastal tourist destinations through the south Asia region for their Christmas holidays. The total death toll exceeded 200 000 throughout the region including tens of thousands dying in Thailand, Sri Lanka and India.

Other famous teletsunamis have occurred, following greater than magnitude 9 earthquakes, in Chile in May 1960 and in Alaska in March 1964. The 1960 Chile tsunami radiated across the Pacific Ocean causing devastation in Hawaii approximately 15 hours, and in Japan 22 hours, after the earthquake. The March 1964 Alaskan earthquake measured 9.4 on the Richter scale and sent large tsunamis into the Pacific Ocean that impacted the west coast of Alaska, Canada, United States, South America, Hawaii and Japan (Dudley and Lee, 1998).

Submarine landslides also generate devastating tsunamis. Often these tsunamis are more localised with the largest waves focused onto a section of coast. These tsunamis have also been known to be very large. The Aitape tsunami in 1998 focused its largest waves onto a section of coast approximately 10 km in length. At Sissano Lagoon, only three waves impacted the shore. The first and second of these waves were between 10 and 15 m in height at the shore (Fig. 5.2). The waves were still at least 5 m high 400 m inland after they had traversed the beach, penetrated a stand of Coconut and Casuarina trees along the seaward side of the barrier, and then crossed a grassy field (Kawata *et al.* 1999). Landslide-generated tsunamis have not been known to spread out across ocean basins like some earthquake tsunamis.

While they are typically more localised, some of the largest tsunamis ever known in historical times were generated by both submarine and subaerial landslides. The largest known tsunami occurred in Lituya Bay, Alaska in July 1958 following an earthquake that caused an avalanche to plunge into the waters of the fiord. A wave 30–50 m high roared down the bay towards the open ocean outpacing several boats attempting to flee, and also propelling one boat with people on board over trees standing at least 30 m high. The wave also ran up the opposite side of the bay to a height of 524 m stripping all of the vegetation from the slope to this elevation. Other localised landslide tsunamis have occurred numerous times. The 1994 Skagway, Alaska tsunami was generated when a railroad dock, heavily laden with soil and rock, and pile driving equipment, collapsed when the sediment supporting the dock slid into deep water.

Figure 5.2. Devastation of tsunami at Sissano Lagoon, July 1998. Note a steel bucket in a tree (arrowed), marking the minimum height of the tsunami wave as it travelled across the beach. Stumps in the ground show the direction of tsunami travel – they supported a wooden house prior to the event.

A 7–8 m high wave washed across the harbour destroying buildings, boats and scouring the harbour floor a further 10 m deeper than its original depth (Dudley and Lee, 1998). In 1999, a relatively small subaerial landslide plunged into the sea from an island in the Marquesas Group in the central Pacific Ocean. The landslide generated a 5 m high tsunami that was witnessed by a school head-master whilst teaching a class. As the wave was rushing towards shore, he was able to evacuate the school children from the classroom through windows just before the wave crashed into the room floating desks and chairs.

It is hypothesised that huge or megatsunamis have been generated in the geological past by the collapse of large chunks from the sides of volcanic islands. Mapping of the sea floor around the Hawaiian Islands revealed extensive areas of slumps and debris flows that emanated from the flanks of Hawaii's volcanoes. Volcanic debris in one slide is 80 m thick and covers an area approximately 20 km wide and 35 km long. The total extent of deposits surrounding the Hawaiian islands come from around 68 major landslides and cover a 20 000 km^2 area of sea floor. Future slides of this scale could generate very large tsunamis that would run up potentially hundreds of metres on nearby populated islands. A similar scenario is predicted for the Canary Islands in the Atlantic Ocean where landslides could generate tsunamis that would travel thousands of kilometres and inundate substantial areas of the US east coast.

Tsunamis have been responsible for nearly a quarter of people killed during volcanic eruptions (Dudley and Lee, 1998). Historically, there have been 92 major tsunamis generated by volcanic eruptions. The tsunamis result from pyroclastic flows entering the water, submarine explosions, earthquakes associated with the eruption, caldera collapse or volcanic landslides. One of the best known tsunamis associated with a volcano occurred during the Krakatau eruption of 1883. Numerous tsunamis were produced, probably by pyroclastic flows entering the sea, the day before the major eruption. The highest of these tsunamis was about 10 m. At 9.58 am on the 27th August 1883, the major explosion of Krakatau occurred generating a 30 m high tsunami that killed many thousands of people. The entire population of one island approximately 13 km away from the centre of the eruption was killed. In all, nearly 40 000 people died due to the series of tsunamis that occurred from the afternoon of the 26th August till approximately midday on the 27th. Tsunamis from this eruption rounded the Cape of Good Hope and entered the Atlantic Ocean to be recorded as small waves in France. Strange tidal oscillations also occurred in Sydney Harbour following the eruption. Whether these were part of a tsunami that ran along the east coast of Australia, probably south to north, or were due to the atmospheric shock wave that circled the globe following the eruption, is not known. A small tsunami was even observed on Lake Taupo on the north island of New Zealand, which being unconnected to the sea must have been generated by the shock wave. It is uncertain which elements of the eruption caused the major tsunami along the coasts of Java and the Sunda Strait. The tsunami could have resulted from the collapse of the caldera, pyroclastic flows entering the sea or submarine explosions, or indeed a combination of all of these may have been responsible.

Meteotsunamis

Meteorological tsunamis are long-period waves generated by either shock waves in the atmosphere or sudden changes in atmospheric pressure associated with squall lines and/or tropical cyclones. They have been recognised the world over and are given a variety of local names such as 'Rissaga' (Mediterranean Sea), 'Abiki' and 'Yota' (Japan), 'Marubbio' (Sicily), 'Seebir' (Baltic Sea) and 'Stigazzi' (Gulf of Fiume). The absence of a seismic generating source suggests to some that they are not true tsunamis, but their physical form and behaviour is identical. Meteotsunamis have periods of many minutes (from several to tens) and will build in height as they approach the shore. Their formation is thought to be related to the formation of an inverted barometer wave on the ocean that can resonate (through Proudman resonance) and amplify in height. This occurs when the atmospheric pressure jump and coupled ocean wave travel at a certain velocity into a certain depth of water. Donn and Balachandran (1969) suggest that the ocean wave will increase in height according to the following equation;

$$v^2/v^2 - c^2 \tag{5.2}$$

where c = the speed of the atmospheric disturbance and v = the water wave speed ($gh^{0.5}$).

The mechanism described in equation (5.2) requires a matching of the phase velocity of the air and water disturbances. A fall of 1 hPa (hPa = mb) in air pressure will result in a corresponding rise in the ocean surface of 1 cm. But this size ocean wave can amplify by 20 times or more (Donn and Balachandran, 1969) in the open ocean. Meteotsunamis have been known to reach 3 m height in the Balearic Islands (Western Mediterranean) (Rabinovich and Monserrat, 1996).

It is also possible that large volcanic eruptions can generate meteotsunamis through the propagation of an atmospheric shock wave. Lowe and deLange (2000) postulated that the AD 200 Taupo eruption in New Zealand may have generated a worldwide meteotsunami, or volcano–meteotsunami.

Modern tsunami impacts on coasts

Tsunamis can both deposit sediments on shore and perform substantial erosion of the coastal landscape (Dawson, 1994). Depositional imprints of tsunamis are often in the form of sheets of sediment that taper landwards from the shore (Dawson and Shi, 2000). Boulders too can be transported and deposited as fields, ridges or trains. Transported boulders have been observed following tsunami impacts in Japan (Sato *et al.* 1995), New Guinea and Java (Dawson *et al.*, 1996).

One of the first observations of the impact of a tsunami on the coastal landscape was following the 1960 Chilean earthquake-generated tsunami (Wright and Mella, 1963). Sand and silt were transported inland and deposited as a thin (2 cm) layer over the ground surface. Other similar deposits have since been observed in several locations following tsunami events, particularly in Indonesia (Minoura *et al.*, 1997; Shi *et al.*, 1995; Yeh *et al.*, 1993) and Japan (Sato *et al.*, 1995). The 1992 Flores Island Indonesia event deposited coral clasts as well as a sand sheet up to 50 cm thick that tapered in thickness landwards. The sheet extended inland for 150 m and the source of the sediment appeared to be the beach and coastal foreshore which suffered severe erosion (Yeh *et al.*, 1993). The 1998 Aitape (Papua New Guinea, PNG) tsunami left a sand sheet up to 1 m thick that extended inland for 675 m (USGS, 1998; McSaveney *et al.*, 2000). The sand sheet consisted of normally graded sediments (coarser-grained at the base becoming finer-grained

Figure 5.3. Sedimentary unit deposited during the December 26th 2004 Asian tsunami in Tamil Nadu, India.

towards the top) with the only sedimentary structures consisting of faint horizontal stratification (layers) in the upper part of the deposit. The 2004 south Asian tsunami deposited planar (horizontal) beds of normally graded sands (Fig. 5.3) that tapered in thickness landwards from 40 cm near the beach to less than 10 cm thick over 300 m inland along the southeast Indian coast. At least six separate horizontally bedded units of sand can be seen in Fig. 5.3 corresponding to the six prominent waves that impacted the coast here.

Erosional features

Erosional features, mainly within barrier and beach sands, have also been observed following a number of recent tsunamis. The sand barrier at Sissano Lagoon, showed two main forms of erosion following the 1998 Aitape tsunami (Figs. 5.4 and 5.5). The rear of the barrier was scarped to a height of 1–1.5 m for a distance of 100–150 m alongshore and below the scarp, scour holes developed ranging in size from 10 m to approximately 1 m diameter and up to 1.5 m deep. Some of the scour holes occurred on the lee, or lagoon side, of trees. These were generally the narrowest diameter scour holes (Fig. 5.4), whereas the

Figure 5.4. Narrow scour pits developed behind trees, Sissano Lagoon. Waves travelled from right to left and where the water flow separated from the ground surface at the back of the barrier, roller-like vortices likely developed to form the longshore scarping (as seen in the background) and scour pits on the lee side of trees.

Figure 5.5. A broad scour pit developed in the absence of trees at the base of a long-shore scarping back barrier, Sissano Lagoon.

larger diameter scours occurred away from trees (Fig 5.5). The form and location of these erosional features suggests that vortices developed within the tsunami flow. The long scarp was probably developed by a horizontal 'roller' type vortice, whereas the scour pits appear to have developed by vertical vortices occurring both within the flow, and also as a function of flow separation around trees where eddies developed within a zone of lower pressure on the lee side of the trees.

Sato *et al.* (1995) also report scour holes between 0.5 and 1.5 m deep in a gravel terrace following the 1993 SW Hokkaido tsunami. The tsunami scour holes developed where a 50 m length of breakwater collapsed concentrating the outgoing tsunami through a narrow opening. About 1500 m^3 of gravel from a Holocene coastal gravel terrace was eroded to produce the bowl shaped depression. Other erosional features were noted by Sato *et al.* (1995) on Okushiri Island. Here the tsunami eroded grooves several metres in diameter trending parallel to the shore. These grooves occurred in two zones both oriented parallel to shore with a maximum depth of 40 cm. They were located about 20–40 m inland of a shore protection wall. One of the groups of grooves formed as water swept over the wall, presumably due to turbulence generated as the tsunami flow separated from the wall itself.

PALAEOTSUNAMI

Atwater (1987) undertook one of the earliest studies of a prehistoric tsunamic deposit. He noted a small forest of dead trees behind a coastal barrier on the coast of Washington State, USA (Atwater and Moore, 1992), where the tree bases are now standing within the intertidal zone. These trees do not normally grow within this zone and Atwater hypothesised that the trees had been drowned after subsidence of the coastal land most likely during an earthquake. He also noted a sand layer in the sediments surrounding the trees. The sand layer occurred within much finer-grained sediments and it became clear that the sand had been deposited from a marine inundation into an area not normally subject to encroachment by the sea. The Cascadia subduction zone, where the Pacific tectonic plate is being subducted below the North American tectonic plate occurs offshore. Such a geological structure is very likely to have produced large earthquakes. Atwater concluded that the sand layer was deposited by a tsunami generated by an earthquake with its epicentre in the Cascadia subduction zone. He was able to determine the age of the earthquake, and tsunami, by analysing the chronology of the drowned trees through radiocarbon dating and tree ring analysis. Interestingly, nearly ten years after Atwater published his study, Satake *et al.* (1996) showed that the same tsunami probably travelled thousands of kilometres across the North Pacific Ocean to strike Japan killing many people. The origin of this tsunami in Japan had previously remained a mystery for it was not associated with any known earthquake felt in Japan. The Japanese historical record was able to show that it had occurred on the 26th January 1700 (Satake *et al.*, 1996), a date that is bracketed by the chronological determinations for the northwest USA tsunami by Atwater. This study is significant for it shows that palaeotsunamis can leave signatures (both historical and presumably geological) of their inundation on coasts far removed from the tsunami-generating source.

Sand sheets

Sand sheets extending inland from the shore are the most common form of evidence used to identify palaeotsunami inundations. Tsunami laid sand sheets vary in thickness from a few centimetres to approximately a metre. They can be both continuous and discontinuous in their extent inland depending upon variations in the onshore local topography, and they can extend inland from tens of metres up to 2 km. Goff *et al.* (1998, 2001) present 16 diagnostic criteria of tsunami laid sand sheets (Table 5.1).

There appear to be minor differences between sand sheets deposited by storm surge and waves (Nanayama *et al.*, 2000). Both wave forms, however, can deposit sand sheets extending hundreds of metres inland (see, for example, the sand sheet deposited by Cyclone Vance in Chapter 4). Goff *et al.* (2004) compared a tsunami and storm deposit on the North Island of New Zealand. They found that the storm deposit was better sorted, coarser grained, didn't decrease in thickness noticeably with distance inland and had a sharp lower boundary whereas the tsunami deposit had an erosional lower contact. Another potential difference between the two is the presence of sedimentary structures showing backwash or the return flow of water to the sea in tsunami events, i.e. fore-set beds showing both the onshore flow and backwash (Fig. 5.6). However, such structures are by no means evident in all tsunami laid sand sheets, especially where backwash did not return over the same area of beach or barrier that the onshore flow transgressed. This was the case at Sissano Lagoon, PNG, where the onshore flow swept across the beach and barrier and into the lagoon; the return flow then travelled back to the sea through the mouth of the lagoon, not back over the barrier. The other potential difference between the two types of sand sheet is the presence, sometimes, of offshore or deeper water foraminifera and diatoms in tsunami deposits. Tsunami can disturb sediments on the sea floor at depths greater than that capable by wind-generated or storm waves because tsunamis are shallow water waves for their entire existence. As a consequence,

Table 5.1 *Diagnostic criteria used to identify tsunami deposits (from Goff et al. 1998, 2001).*

1. Sediments within sheet generally fine inland and upwards. Deposits often rise in altitude with distance inland
2. Each wave can form a distinct sedimentary unit, although this is not often recognised in the sedimentary sequence
3. Distinct upper and lower subunits representing run-up and backwash can be identified, *but* investigation of recent tsunami deposits indicates that there is still considerable uncertainty about when most deposition occurs (during run-up or backwash) and so these subunits may be related to other processes
4. Lower contact is unconformable or erosional
5. Can contain intraclasts of reworked material, but these are not often reported
6. Often associated with loading structures at base of deposit
7. Particle and grain sizes range from boulder layers (up to 750 m^3), to coarse sand to fine mud; however, most deposits are usually recognised as anomalous sand units in peat sequences
8. Generally associated with an increase in abundance of marine to brackish water diatoms, but reworking of estuarine sediments may simply produce the same assemblage. Preservation of frustules can be excellent, although many are often broken
9. Marked changes in foraminifera (and other microfossil) assemblages. Deeper water species are introduced with catastrophic saltwater inundation
10. Increases in concentration of sodium, sulphate, chlorine, calcium and magnesium occur in tsunami deposits relative to under- and overlying sediments; indicates saltwater intrusion
11. Pollen concentrations are lower (diluted in the deposit)
12. Individual shells and shell-rich units are often present
13. Often associated with buried vascular plant material and/or buried shell
14. Shell, wood and less dense debris often found 'rafted' near top of sequence
15. Dating of tsunami sediment is problematic. Best results for dating are from units above and below to 'bracket' the event. Radiocarbon ages often equivocal because of older reworked carbon; dating of introduced marine macrobiota is preferred (and successful). Optical dating (OSL) is the best method available assuming the sediments were exposed to daylight during reworking by the tsunami
16. Often associated with reworked archaeological remains (e.g. middens). In some cases occupation layers are separated by a palaeotsunami deposit

foraminifera and diatoms that live at greater water depths can be incorporated within the tsunami flow and be deposited onshore, whereas storm waves tend to entrain shallower water species.

Sato *et al.* (1995) suggested that the spatial distribution and thickness of tsunami sand sheets depends on the size of the tsunami inundation, in particular tsunami run-up, and the availability of sediment for entrainment. Based

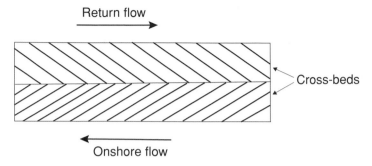

Figure 5.6. Diagrammatic example of sedimentary structures (cross-bedded foresets) showing both onshore and return (seaward) flow.

upon their studies of deposits from two tsunami inundations in Japan (Japan Sea tsunami of 1983 and SW Hokkaido tsunami of 1993), they suggested that tsunami run-ups of greater than 10 m were necessary to produce a continuous sand sheet as opposed to more patchy deposits of sand.

Landward thinning of sand sheets is an important characteristic of tsunami deposits, although the same occurs with storm deposits. But this characteristic at least distinguishes between sediments laid down by marine inundations and those by wind and rivers. The size of the tsunami (wave height at shore and run-up height), along with local onshore topography, will also have some bearing on the distance inland that a sheet will extend. Given a generally flat coastal landscape, tsunamis that extend farther inland will produce more extensive deposits, providing there is an adequate supply of sediment. The tsunami at Sissano Lagoon, PNG, whilst very large, was unable to deposit an extensive sand sheet because of the proximity of the lagoon at the rear of the sand barrier. Dawson *et al.* (1988) found a sand sheet extending inland for over 2 km in east Scotland. The tsunami responsible was thought to have been generated by the Storegga submarine landslide off the coast of Norway approximately 7950 years ago (Fig. 5.7). This was the second of three massive submarine landslides which also left landward thinning layers of sediments within tectonically raised lakes on the east coast of Norway (Bondevik *et al.*, 1997). Bondevik *et al.*, (2003) have also identified the Storegga slide tsunami in the Shetland Islands where there is evidence that the tsunami attained a run-up height of 20–25 m above sea-level at the time. There is evidence for two waves in the tsunami train here. The first ripped up clasts of peat in coastal lakes and left these behind after the wave drained the land as backwash. The second wave incorporated the ripped up clasts into a sand layer that thins inland and decreases in grain size in the same direction (Fig. 5.8).

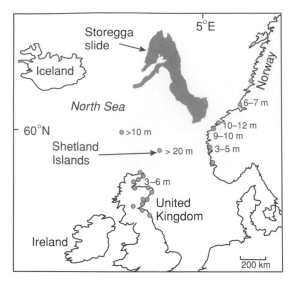

Figure 5.7. Location of Storegga landslide and run-up heights of resultant tsunami in Norway, Scotland and the Shetland Islands (after Bondevik *et al.*, 2003).

Figure 5.8. Storegga landslide-generated tsunami deposit in Sullom Voe, Shetland (photograph courtesy of Dr S. Bondevik, University of Tromsø, Norway). The tsunami deposit is composed of rip-up clasts of peat and pieces of wood interbedded with sand. Bondevik *et al.* (2003) interpret this as a result of two waves with the first wave eroding the peat surface and transporting the rip-up clasts of peat and the backwash leaving the eroded clasts and other organic remains at the surface of the tsunami-laid sand. The second wave buried the clasts in sand.

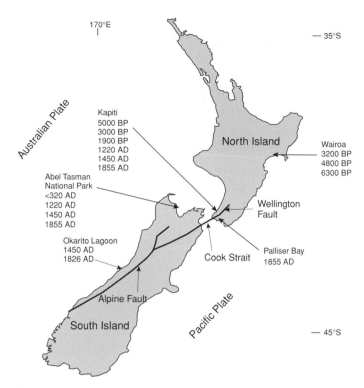

Figure 5.9. New Zealand palaeotsunami deposit sites and chronologies (after Goff *et al.*, 1998, 2001, 2004a).

Sand sheets that extend into back-barrier marshes and lagoons are easily identified as coarser-grained sediment layers within otherwise fine-grained sediments. Grain size within these sheets typically decreases upwards and landwards at any one location. These changes in grain size are an important diagnostic tool for identifying both multiple separate tsunami events and separate waves within the same tsunami event. Each of these scenarios produces a series of upward fining sediment cycles with the base of each new event marked by the start of coarser-grained sediments and the top marked by the finest-grained sediments. The next wave or tsunami event is indicated by the next similar layer. Many studies have identified multiple events and waves within a single event based upon the nature of sediment cycles. For example, Goff *et al.* (1998) identified the AD 1855 Wellington (New Zealand) tsunami generated by a severe earthquake resulting in a 9–10 m high tsunami in Cook Strait and a 3–4 m high tsunami at the entrance to Wellington Harbour (Fig 5.9). Three waves were known to have inundated the coast during this event; the first wave was the largest. Goff *et al.* (1998) found three cycles of upward fining sediment at seven separate sites examined along the coast near Wellington Harbour (Kapiti, Fig. 5.9). Each layer varied in thickness from 10 cm to over 80 cm depending upon the site. The layers were typically marked by clast-supported gravel (i.e. gravel resting against gravel without any intervening sediment such as sand or silt) at their base merging upwards into coarse-grained, then fine-grained sand and silt at the top of the sediment cycle. The sequence changed abruptly into the next overlying layer starting with another unit of clast supported gravel. The same sequence was present in the third or uppermost layer marking the inundation of the coast by the third tsunami. A similar sequence of upward fining sediment cycles was also observed by Shi *et al.* (1995) following the 1992 Flores Island, Indonesia tsunami. Four sediment cycles are present within coastal lake deposits in eastern Norway marking the Storegga landslide-generated tsunami. Each of the tsunami deposit layers is composed of coarse-grained sands and shell grading upwards into finer-grained sediments and then plant fragments (Bondevik *et al.*, 1997).

The change in grain size within successive tsunami sediment cycles is likely to be due to the decrease in energy of the wave as it passes over the coast. The coarsest-grained sediments are transported as traction load being either rolled or bounced along the bottom, whilst the finer-grained sediments are carried in suspension. As the wave energy diminishes both at a location and as it travels inland, the heaviest or coarsest-grained sediment is deposited first followed by the lighter or finer-grained sediment. Such a scenario, however, assumes that there is a variety of different sediment grain sizes available for transportation by the tsunami. Dawson *et al.* (1996), in their study of the 1994 Java tsunami deposit, found no variation in grain size with distance inland. They attributed this to the uniformity of the sediment grain size available for transport at this location.

Sediment cycles associated with multiple tsunami events are usually marked by a significant time gap or hiatus that can be identified either by geological dating of the different layers, or the presence of a buried soil at the top of each event cycle. The soil shows that the land surface following a particular tsunami event had remained stable (not inundated or disturbed) so the upper layer of sediments, at that time, was able to weather and experience pedogenesis (soil formation). Subsequent tsunami inundations then deposit sediments on top of the soil. Separate tsunami events can also be distinguished by sharp erosional contacts between layers. These contacts are abrupt and the base of the overlying sediment unit can often contain 'rip-up' clasts of material derived from the ground surface during passage of the tsunami. McSaveney *et al.* (2000) identified two tsunami events prior to the 1998 event in the sediments near Sissano Lagoon, Papua New Guinea. They suggested that these events may have been associated with the AD 1907 and 1934 tsunamis at this location. The studies of Atwater (1987),

Darienzo and Petersen (1990), Clague and Brobrowsky (1994) and Clague *et al.* (1999), identified six separate tsunami events along the northwest USA and southwest Canadian coasts from buried sand layers separated by fine-grained sediments and peat dating between 300 and 3200 years BP. Twelve layers of sand generated from separate tsunami events have also been identified in a peat bog in Crescent City, California. These tsunamis would have been generated from large earthquakes in the Cascadia subduction zone. Nanayama *et al.* (2003) identified 17 tsunami sand layers separated by peat and volcanic ash on the Japanese island of Hokkaido. These sand sheets extend inland for up to 3 km and were deposited over the last 2000–7000 years giving an average return interval between events of 500 years. Comparisons between the spatial extent of these sheets and those of historically recorded tsunami over the last two centuries, showed that the earthquakes responsible were very large, involving multiple segment ruptures along many hundreds of kilometres of the Kuril Trench where the Pacific tectonic plate converges with Eurasia. Historical tsunamis here have only been generated by earthquakes from single segment ruptures of 100–200 km length, which was deemed to be the normal style of earthquake and tsunami inundation for the region.

Few studies have reported clearly identifiable sedimentary structures within tsunami sand sheets as the majority of deposits display massive sedimentation. However, surveys of the sand sheet deposited by the 1998 Aitape, Papua New Guinea tsunami near Sissano Lagoon reported faint horizontal stratification towards the top of the deposit (USGS, 1998) (Fig. 5.1), and Sato *et al.* (1995) reported cross bedding in the 1993 southwest Hokkaido deposit. Lamination bedding was also reported in other Japanese tsunami deposits (Fujiwara *et al.* 2000). Sometimes load structures (where depressions are made within an underlying deformable unit of sediments) are observed at the base of the deposit (Goff *et al.*, 1998).

Goff *et al.*'s (1998) study of the AD 1855 tsunami deposit near Wellington, New Zealand examined the roundness and sphericity of lithic clasts, as well as the fabric of clasts (i.e. alignment of *a* axes), and the presence and nature of pumice and marine shells. Marine lithic clasts tend to have a flatter disk shape than the rounded gravels of rivers, because marine clasts experience continuous movement up and down the beach with the swash and backwash associated with normal fair weather waves. Clast axis alignment tends to result in the longest axis (*a* axis) aligning parallel or subparallel to the shore or perpendicular to the direction of wave transport. This is because the clast can be more easily transported in a fluid flow by rolling over its shortest axis (*c* axis) which is perpendicular to the *a* axis. Clasts can also become imbricated in fluid flows and in the case of marine inundations the clasts will normally dip seaward or

Figure 5.10. Imbricated clasts of rock deposited inland of the beach during the surge of Tropical Cyclone Vance, Western Australia March 1999. Onshore flow was from left to right.

opposite to the direction of fluid flow. Imbricated coral clasts were present on islands affected by the 1998 Aitape, PNG tsunami. Imbricated clasts can also occur during deposition of gravel by storm waves as was the case in Western Australia during Tropical Cyclone Vance in 1999 (Fig. 5.10).

The presence of diatoms and foraminifera are common in tsunami laid sand sheets. Dawson *et al.*'s (1996) study of diatoms showed that tsunami deposits often have a chaotic mixture of species from different environmental settings including marine and freshwater species. They attributed this to tsunamis transgressing a range of habitats as they travel onshore. Diatoms are often found in a broken state in tsunami deposits with some more resistant species able to survive and remain intact. Freshwater species, having been transported from marshes over a shorter distance, can often be found still intact. The same applies to foraminifera. Dominey-Howes (1996) studied formanifera in tsunami deposits in Greece and Crete (Dominey-Howes *et al.*, 2000). He noted that deeper water species were high in number in the deposits.

Bryant (2001) suggests that tsunami and storm surge deposits will display different diatom and foraminifera assemblages. It is proposed that only larger benthonic (sea floor dwelling) species are transported shoreward as bedload by fair weather waves and deposited onto the beach. The smaller benthonic and planktonic (sea surface dwelling) species are moved offshore by backwash, particularly so during storm waves, and back onshore by winds blowing the surface of the water. The difference between storm and tsunami assemblages is that the former have smaller diatoms and foraminifera with some of the larger species incorporated from the beach, whereas tsunami deposits have many larger species amongst a chaotic mixture of species from different environments including freshwater settings. Hemphill-Haley (1996) has warned that microfossil assemblages cannot by themselves differentiate between tsunami and storm deposits. And Minoura et al. (2000) found quite the opposite sort of assemblage in a study of deposits from the Minoan tsunami generated by the Santorini volcanic eruption ~3600 years BP. This deposit showed very high numbers of planktonic foraminifera species and an absence of species that dominate the modern shore.

Seawater is rich in chlorine (Cl), sodium (Na) and sulphates (SO_4) and such elements are much less common in terrestrial waters. Na, potassium (K), calcium (Ca) and magnesium (Mg) also occur in seawater with the latter two high in marine shells and foraminifera. Geochemical studies of sediments thought to have been deposited by tsunami are frequently high in these elements. Minoura and Nakaya (1994) found such trends in their study of tsunami sediments in a Japanese coastal lake and Minoura and Nakaya (1991) found similar levels of these elements in sediments deposited by the AD 1993 Japan Sea tsunami. Goff et al. (2004) also analysed geochemical signatures in the sediments of Okarito Lagoon, New Zealand (Fig. 5.9). Here they found that iron (Fe), sulphur (S), titanium (Ti), strontium (Sr), barium (Ba) and Na concentrations increased markedly in a core which at the same level showed a distinct change in sediment grain size. Goff et al. (2000, 2004) suggest that these increases in the chemical composition of the sediments reflect the influence of saline water (tsunami inundation) into the normally brackish to freshwater lagoon.

Identification of tsunami deposits is best achieved when all of these criteria are employed within a study. Goff et al.'s (1998) study of the AD 1855 tsunami deposit near Wellington, New Zealand examined the sediment textural characteristics, the diatom assemblage, the roundness and sphericity of lithic clasts, the fabric of clasts (i.e. alignment of a axes), and the presence and nature of pumice and marine shells. Likewise, Goff and Chagué-Goff's (1999) study incorporated these same elements plus geochemistry and radiocarbon and ^{137}Cs analysis to determine the age of the tsunami events in Abel Tasman National Park, New Zealand.

Figure 5.11. Boulder ridge near NW Cape, Western Australia. Note the imbrication of clasts and the person standing in the background for scale.

Boulder deposits

Many coasts worldwide display large lithic boulders on rock platforms, and sometimes on top of cliffs tens of metres high. Often these boulders are clustered into groups that contain from tens to many hundreds of individual clasts, some of which can weigh over 100 tonnes, and have a axes as long as 6 m (Fig 5.11). Boulders within these groups are also often imbricated and have aligned a axes. Sometimes they may contain barnacles and/or oysters on their surface which suggests they originated in the near-shore environment. The same is true when the boulders display clear imbrication and axis alignment. Often comparisons with accumulations of core stones due to deep weathering bear no resemblance to the depositional fabric of the imbricated boulder fields, nor do rock falls from cliff collapse (Nott, 1997). However, a boulder located at the coast is not necessarily evidence that it arrived at its present position following transport by waves. The possibility of rock falls and deep weathering always needs to be addressed. Where these processes can be eliminated, and imbrication and alignment are present, there is little doubt that those boulders have experienced wave transport. But the question remains as to what type of wave was likely to have been responsible.

Storm waves can also transport and deposit boulders. One of the most important differences between storm and tsunami waves is their respective abilities to transport boulders at the shore. Tsunamis are able to surge across the shoreline without rapidly reducing their velocity and energy, whereas storm waves break and begin to rapidly dissipate energy in water depths at least 1.2 times the wave height. Hence, a 5 m high storm wave will break and begin to dissipate energy at a water depth of 6 m. In many situations this means that the storm wave will break many hundreds of metres and sometimes kilometres offshore, depending upon the offshore bathymetry. Storm waves can maintain substantial energy close to the shore where the water immediately offshore is deep. It is in these situations that storms have the greatest likelihood of transporting large lithic boulders along or onto the shore. Large waves can strike the shore, therefore, if the adjacent waters are sufficiently deep to sustain that wave. Such situations occur adjacent to steep sea cliffs (Williams and Hall, 2004). Deposition of coral boulders atop steep cliffs, tens of metres high, has been noted during rare conditions, e.g. Niue Island in the Pacific during the passage of Tropical Cyclone Ofa in 1990 and again during Tropical Cyclone Heta in 2004. Alternatively, large storm waves can occur close to shore if accompanied by a storm surge, which can temporarily increase the depth of the normal inshore water by up to several metres (Table 5.2). In these situations the size of storm waves possible at the shore is determined by the height of the storm surge and this is in turn a function of a number of specific characteristics (see Chapter 4) of the tropical cyclone involved. Differentiating between boulders deposited by storm waves and tsunamis is not easy. Table 5.2 describes the different processes occurring during transport of coastal boulders by these two wave types whereas Table 5.3 outlines the sequence of events required to be ascertained when attempting to determine which wave type was responsible for boulder deposition.

Studies of coastal boulder movements

One of the earliest studies of boulder movement by waves was undertaken by Sussmilch (1912). Sussmilch noted the emplacement of a boulder measuring approximately $6 \times 5 \times 3$ m^3 on a shore platform near Bondi Beach, Sydney, Australia, following a temperate storm the previous night. The boulder, now known as Mermaids Rock, stands as an isolated feature on the shore platform. Sussmilch suggested that the rock had been lifted vertically approximately 3 m after it was dislodged from the seaward face of the shore platform and transported approximately 20 m across the platform. Other smaller boulders were also transported during the event, but these also appear to have been left as isolated features, not displaying any imbrication. It is uncertain whether

Table 5.2 *Conditions required for boulder transport by storm waves and tsunamis*

Situation/process	Storm	Tsunami
Wave height at shore	Breaks and dissipates energy in water depths 1.2 times wave height	Not restricted by offshore water depth, can surge across dry shore
Velocity at shore	$gh^{0.5}$ (0.25 that of tsunami) only when wave breaks at shore, less when wave breaks offshore	$2gh$ (4 times greater than storm)
Work performance	Wave height needs to be approximately 4 times larger than tsunami to move given size block	One-quarter height of storm wave needed
Boulder deposition on cliff tops	Must have very deep water immediately offshore, lower velocity restricts ability of wave (fluid flow, not spray and projected water) to climb cliffs	Can climb cliffs at least 5 times wave height at cliff base irrespective of water depths offshore
Need for accompanying events	Where offshore waters are shallow, large waves need storm surge to support breaking at shore, surge heights have physical constraints (often restricted to less than 4–5 m)	Do not need storm surge
Source of waves	Largest waves restricted to mid-latitude westerly belts and tropical cyclone prone regions	Most common near convergent tectonic plate boundaries, but possibility of submarine landslides, bolide impact and meteotsunamis (in this sense probably have more extensive source areas when considering geological time frame)

the large boulder was overturned during transport, slid or was projected to its present position.

Noormets *et al.* (2002) undertook a study of boulder movement on the shore platform at Sunset Beach, Oahu (Hawaii), a location renowned by surfers for big waves. They studied a series of aerial photographs of the site dating from AD 1928 to 1996 and noted the position of boulders on the platform both before and after known tsunamis and big storm wave events. They observed that the largest lithic clast (96 t) was most likely emplaced as a solitary feature during the 1946 Aleutian tsunami which had a wave run-up of between 3.3 and 10.7 m at various locations on the island of Oahu, but 4.3 m near Waimea Bay (closest

Table 5.3 *Suggested sequence of events required to determine whether boulders were transported by tsunamis or storm waves*

Best evidence

Direct observation of event, or observations and measurements before and after event

Theoretical evidence

1. Establish that boulders have been transported by waves (imbrication, axis alignment, clast size grading, presence of marine shells adhered to boulder surface)
2. Assess whether boulders have been overturned during transport, slid to present position, or been blasted or tossed by wave impact on rock (shore platform) face
3. Determine likely pre-transport source of boulders (subaerial, submarine, joint bounded blocks, cliff face)
4. Determine relationship between boulder size and wave height needed for transport (also depends on pre-transport position of boulder)
5. Determine observed and theoretical storm wave climate (using numerical wave models)
6. Determine maximum possible storm surge (tropical and temperate coasts)
7. Determine offshore bathymetry and distance from shore of maximum breaking storm wave height
8. Determine whether tsunami possible (not just likely) from any source
9. Determine, where possible, age of deposition

---

location to study site). The boulder has moved twice since its initial emplacement, both times by approximately 30 m in different directions. Noormets *et al.* (2002) suggested that the first of these movements could have been due to a subsequent tsunami and the latter movement was most likely by very large storm waves in 1969 (the 'Big Wednesday' event). Any possible movements of this boulder, or other smaller ones, on the shore platform during the January 1998 'Biggest Wednesday' swell in the same area, however, are not mentioned in the study. The waves from the 'Biggest Wednesday' swell are regarded by surfers in Hawaii as the biggest waves seen in modern times in this region, and were probably larger than those occurring during the 1969 event.

Nott (2004) studied a 1 km long boulder ridge, at the rear of a beach, near the entrance to Exmouth Gulf, Western Australia to see if one of the most powerful tropical cyclones (TC Vance) to strike the area had any impact on the ridge. The ridge is composed entirely of Quaternary sandstone 'beach-rock' clasts measuring up to $3.0 \times 2.1 \times 0.7$ m^3 (Fig. 5.11). The clasts are imbricated with seaward dips, and the majority are orientated with their *a* axes parallel or subparallel to shore. The ridge was completely overwashed by the inundation generated by TC Vance.

Surveys before and after TC Vance showed that this ridge was not modified by the waves and surge generated by this cyclone. This was despite the fact that the ridge was fully exposed to the near maximum force of the winds and waves (this location was approximately 10 km away from the eye of TC Vance where the winds were estimated at over 300 km h^{-1}). No new boulders appeared to have been added to or removed from the ridge during TC Vance. The boulders in the ridge were originally derived from an intertidal platform of beach rock that occurs adjacent to the entire length of the ridge. In order for the waves generated during TC Vance to have supplied new boulders to the ridge, they needed to first excavate blocks from the platform adjacent to the ridge; there were no isolated boulders lying on the beach between the platform and ridge.

Boulders on the ridge are encrusted with oysters which were radiocarbon dated and returned ages between AD 670 and 1820. The ages represent the time of death of the oysters and their death would have been caused by the excavation of the clast and its removal from the intertidal position. The radiocarbon chronology shows a variety of ages suggesting that the ridge accumulated over a number of events, i.e. each new event probably excavates and deposits new boulders to the ridge. The youngest ages came from small boulders encrusted with oysters at the back of the ridge. These smaller boulders (a axis = 28 cm) could have been excavated from the shore platform and carried over the ridge by a tropical cyclone as they are small enough to have been entrained by storm waves or tsunamis. The dated sample of shell from 8 m elevation in the sand dunes behind the ridge, also returned a young age, and likewise it is equally possible for it to have been deposited during a storm or tsunami. The other ages came from oyster encrusted boulders that appear too large to have been entrained by storm waves and suggest they were deposited by separate tsunami events over the past approximately 1000 years.

A number of other studies of wave deposited boulders are also published in the scientific literature. These include, to name but a few, Scheffers (2002), Williams and Hall (2004) and Young *et al.* (1996).

Determining the type of wave responsible for boulder movements: a theoretical approach

Because boulder movements by waves are relatively rare, it is not always possible, to have either direct observations or indirect ones through aerial photos, to ascertain the relative roles of storm waves and tsunamis. Another method is to mathematically model the forces required to transport boulders from different environmental settings by both storm waves and tsunamis.

Hydrodynamic equations that relate the forces necessary to entrain and transport boulders of various sizes, shapes and densities, by waves were first developed by Nott (1997), and later more sophisticated versions that consider the pre-transport environment of the boulder were developed (Nott, 2003). These equations relate four forces; drag, lift, inertia and restraint, incorporating the effect of buoyancy, acting upon a boulder of given shape to the velocity of the breaking wave. Expressed mathematically, the first three forces must exceed or equal the force of restraint in order to initiate transport of a boulder (involving overturning), such that

$$F_d + F_l + F_m \geq F_r \tag{5.3}$$

Where:

$$F_d \ (\text{drag force moment}) = \{0.5\rho_w \, C_d \, (ac) \, u^2\} \, c/2 \tag{5.4}$$

$$F_l \ (\text{lift force moment}) = \{0.5 \, \rho_w \, C_l \, (bc) \, u^2\} \, b/2 \tag{5.5}$$

$$F_m \ (\text{intertia force}) = \rho_w \, C_m \, (abc)\ddot{u} \tag{5.6}$$

$$F_r \ (\text{restraining force moment}) = (\rho_s - \rho_w) \, (abc) \, g \, b/2 \tag{5.7}$$

with ρ_w = density of water at 1.02 g ml^{-1}, ρ_s = density of boulder at 2.4 g cm^{-3}, C_d = coefficient of drag = 2, C_l = coefficient of lift = 0.178, g = gravitational constant, $\ddot{u}$ = instantaneous flow acceleration, u = flow velocity/bore celerity, a = A axis of boulder, b = B axis of boulder, c = C axis of boulder.

Combining these forces together, as expressed in equation (5.3), and incorporating the velocity for storm waves at breaking point as $gh^{0.5}$ and tsunami as $2gh$, gives three different equations depending upon the pre-transport setting of the boulder. For submerged (submarine) boulders where,

$$F_d + F_l \geq F_r \tag{5.8}$$

The equation relating wave height to boulder size for tsunami is

$$H_t \geq 0.25 \, (\rho_s - \rho_w)/\rho_w \, 2a/[C_d \, (ac/b^2) + C_l] \tag{5.9}$$

where H_t = height of tsunami, and, for storm waves

$$H_s \geq (\rho_s - \rho_w)/\rho_w \, 2a/[C_d \, (ac/b^2) + C_l] \tag{5.10}$$

where H_s = height of storm wave at breaking point.

For subaerial boulders (where the boulder is exposed on dry land such as a shore platform) inertia force must be incorporated into an equation to describe the impact of a wave upon that boulder; in a submerged position the boulder is buttressed by the water and the inertia force is absorbed by the boulder. A subaerial boulder will be transported when,

$$F_d + F_l \geq F_r - F_m \tag{5.11}$$

Incorporating equations (5.4)–(5.7) into equation (5.11) gives

$$H_t \geq 0.25 \, (\rho_s - \rho_w)/\rho_w \cdot 2a - C_m \, (a/b)(\ddot{u}/g)/[C_d \, (ac/b^2) + C_l] \qquad (5.12)$$

for tsunamis where H_t = height of tsunami, and,

$$H_s \geq (\rho_s - \rho_w)/\rho_w \cdot 2a - 4C_m \, (a/b)(\ddot{u}/g)/[(C_d \, (ac/b^2) + C_l] \qquad (5.13)$$

For storm waves where H_s = height of storm wave at breaking point.

To initiate motion of a joint bounded block (where the block is located in a shore platform but is separated from other blocks by joints), the lift force must overcome the force of restraint less buoyancy providing the block has weathered completely free from its substrate. The equation for tsunamis in this situation is

$$H_t \geq [0.25 \, (\rho_s - \rho_w)/\rho_w \cdot a]/C_l \qquad (5.14)$$

where H_t = height of tsunami, and for storm waves,

$$H_s \geq [(\rho_s - \rho_w)/\rho_w \cdot a]/C_l \qquad (5.15)$$

where H_s = height of storm wave at breaking point.

Nott (2004) applied these equations to help resolve the origin of the boulders composing the ridge at North West Cape, Western Australia. Numerical storm surge models show that the highest storm surge attainable here is about 2.5 m. The crest of the ridge sits at approximately 3 m above Australian Height Datum (AHD), which is about mid-tide level. The tide range at Exmouth is around 2 m, so even at the highest tide the surge could only be 0.5 m above the crest of the ridge during the most extreme surge. The storm waves required to move the larger boulders onto this ridge range between 4 and 6 m for boulders in either a submerged or subaerial position prior to transport (Table 5.4). Storm waves of much greater heights (>10 m) are required if the boulders were positioned as joint bounded blocks prior to transport, as many of these boulders probably were, for they have been derived from the 'beach rock' shore platform seaward of the ridge. It seems unlikely, therefore, that even the most extreme intensity tropical cyclones are able to produce a surge of sufficient height to support the necessary height storm waves to transport and deposit many of the boulders on this ridge.

Other forms of evidence for palaeotsunami

Large waves can also leave other fingerprints in the coastal landscape besides sand sheets and boulder accumulations. These include deposits of shells and corals atop high-elevation headlands, fluid carved channels also on head-

Table 5.4 *Wave sizes required to transport boulders from various pre-transport settings, Boulder Ridge, Exmouth, Western Australia (from Nott 2004)*

a axis	b axis	c axis	Vol	T (sm)	S (sm)	T (sa)	S (sa)	T (lift)	S (lift)
3	2.1	0.7	4.4	1.4	6	1.3	5.3	4.5	17.9
2.4	1.8	0.7	3.0	1.0	4	1.1	4.2	3.6	14.3
2.8	1.8	0.6	3.0	1.2	5	1.2	4.9	4.2	16.7
2.2	1.5	0.5	1.7	1.0	4	0.9	3.6	3.3	13.1
2	1.2	0.7	1.7	0.5	2	0.9	3.6	3.0	12.0

Boulder axes (a, b and c) measured in metres, vol = volume of boulder (m^3), T = tsunami, S = storm wave, (sm) = submerged boulder, (sa) = subaerial boulder, (lift) = boulder lifted from joint bounded block position through lift force only. Density of boulders measured at 2.1 gm m^{-3}. Wave equations presented in Nott (2003).

lands along with displacement of layers of cap rock strata, and deposits of poorly mixed sediments (diamicts) including sand, shells, corals and lithic fragments several centimetres in diameter (Bryant *et al.*, 1992, 1997; Bryant and Nott, 2001).

Bedrock erosional features along the coast remain a more equivocal form of evidence for past tsunamis. Their similarity to such forms in fluvial and glacio-fluvial environments, and the fact that they often cut across structures and bedding planes, suggests that flowing water may be responsible. However, the precise physical mechanisms for the formation of these features in coastal environments remain to be determined. Until recently their presence in the coastal environment had not been discussed in the literature (Bryant and Young, 1996). Bedrock sculptured features in fluvial environments most likely form during extreme flow velocity conditions, and the same is likely in the coastal environment. The critical question is how rapidly can they form? Certainly, some of these forms were present in the sand barrier at Sissano Lagoon following the 10–15 m high tsunami of July 1998. Here, depressions ranging in size from 0.5 to 1.0 m depth and 0.5 to 8 m diameter were eroded into the rear of the barrier. But these features formed in unconsolidated sands, not bedrock. At the moment the origin of the coastal bedrock erosional features is far from resolved; numerical modelling may help to resolve this situation.

A study of marine inundations at Cape Leveque, Western Australia found all of the above mentioned signatures of palaeowave deposits, in this case most likely attributable to tsunamis (Nott and Bryant, 2003). Cape Leveque is about 200 km north of Broome on Australia's northwest coast (Fig. 5.12). Evidence for large marine inundations here include fields of imbricated boulders, sedimentary

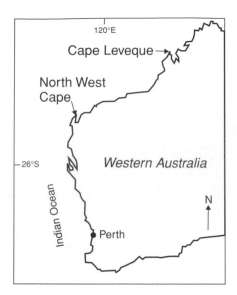

Figure 5.12. Location map showing North West Cape and Cape Leveque, Western Australia.

deposits with marine faunal inclusions atop a 30 m high headland, erosion (plucking) of bedrock and subsequent overturning of the dislodged blocks atop the same headland, and fluvial-like channels, cavettos (scoop-shaped features in channel walls) and pot holes carved into regolith across the headland.

The boulder field at Cape Leveque includes clasts up to 5.5 m (*a* axis) and 4 m (*b* axis). They have been transported across the shore platform and deposited at the base of the headland. Most boulders are imbricated with seaward dips and have their *a* axes aligned parallel to shore. The imbrication and alignment record the refraction of the waves around an island 300 m offshore. The shore parallel alignment of the boulders on two sides of the headland differs by about 90°. Boulders on the west side dip west, those on the north side dip north.

The headland at Cape Leveque is composed of deeply weathered, saprolitic, horizontally bedded Cretaceous siltstone and sandstone. The base of weathering is at the level of the shore platform, which consists of the unweathered part of the rock unit. Many boulders at the headland base are unweathered Cretaceous strata. This suggests they must have come from the shore platform, not from collapse of the headland cliff. The boulders accumulated at the headland's base where the transporting velocity of the flow was checked. Seaward of the boulder field the shore platforms are largely devoid of boulders. In places joint-bounded blocks have been partly lifted and these blocks now rest with one edge, usually the landward edge, upon the shore platform and the remainder of the block sitting within the cavity from which it came. These blocks are often much larger

Figure 5.13. Overturned slabs of the rock 'porcellanite' on top of a 25 m high headland, Cape Leveque, Western Australia. These rocks come from a bed of rock that was originally horizontally bedded. The layering now seen to be orientated in a number of directions was all originally horizontal.

than those plucked and transported into the main boulder field. The velocity of the flow responsible must not have been sufficient to completely entrain these blocks. Possibly the boulder field resulted from more than one event, perhaps a series of them.

Waves also appear to have overtopped the headland above the boulder field. A layer of 'porcellanite' sits 20–25 m above sea level near the crest of the headland (porcellanite is a local term for a silicious rock that forms within a saprolite profile by silica precipitation during the weathering process). The porcellanite here is 0.3–1 m thick and like the original Cretaceous strata is horizontally bedded. It forms horizontally bedded caprock on hills and mesas of Cretaceous strata throughout tropical northern Australia. But today this horizontal stratum is fractured and individual blocks are overturned or standing on end partially buried in the weathered red soil (Fig. 5.13). It is difficult to conceive how pedogenic processes could account for the present orientation of the blocks, for there is no evidence for vertical shearing within this soil as a result of the expansion and contraction of clays. Nor is there evidence for karst processes occurring within the weathering profile that could cause the porcellanite layer to slump. It is more likely that these rocks were overturned by fluid flow over the headland crest.

Other evidence supports the hypothesis that the porcellanite rocks atop the headland were overturned by fluid flow. A channel network deepens and widens eastwards across the headland, suggesting that waves over-ran the headland from the west-northwest. The main channel on the northern side of the headland is 5–6 m across and as deep as 3 m. Minor tributaries feed into the channel head. Clasts of the porcellanite lie within the drainage channel. Various species of marine shells (*Melo Amphora, Phasianell sp.*), oysters and coral fragments are scattered across the headland. Dunes of sediment weathered from the Cretaceous strata also contain gravel derived from the shore platforms as well as shell and coral. These dunes sit on a section of the headland about 30 m above sea level (a.s.l.), above the general level of the dislodged porcellanite layer and the wave overwash channels. Farther south along the western edge of the headland, drainage seems to have been toward the southwest. Here a network of channels deepen up to 12 m and widen in the direction of flow. The channels are carved into the weathered but still-consolidated Cretaceous strata (Fig. 5.6). The channel walls have well-developed cavettos and potholes a few metres deep are numerous. In the same part of the headland, vegetation has been removed, and an obvious trim line runs south along the west side of the headland. Today no vegetation grows west (seaward) of this trim line. Yet clearly vegetation used to be here, for abundant iron-indurated root casts occur in the upper soil layer.

Wave-transported sediment at the back of the beach, 400 m northeast of the headland, includes dunes up to 8 m high. These dunes contain abundant angular to rounded gravel of Cretaceous sandstone and porcellanite from the headland, together with a variety of species of shells, including oysters, and coral fragments. The evidence at Cape Leveque suggests the following sequence of events:

(1) waves ran up over the headland to 20–30 m a.s.l., stripping several metres of orange sandy regolith and maybe also disrupting the porcellanite layer creating boulders of this material;

(2) partial burial of these boulders, either in this event or a later one, and deposition of shell and gravel-rich sand dunes at 30 m a.s.l;

(3) channels cut below the level of the high-gravel and shell-rich dune. The erosion of these channels may have been contemporaneous with deposition of sediments on the headland.

The height of the headland and modifications to it by waves suggest tsunamis to be the most likely cause. This site has experienced tsunamis with run-ups as high as 6 m above mean sea level in AD 1977. A tsunami of this height is too small to have caused the impact to the headland, suggesting that much larger tsunamis have occurred here in prehistoric times. Whether these larger tsunamis were caused by earthquakes in Indonesia as occurred in 1977 remains to be determind. If earthquakes were responsible then they must have been of considerable size compared to those that have generated tsunamis in western Australia historically.

Conclusion

Tsunamis are a spectacular natural hazard that promote both fear and fascination the world over. They are a widespread hazard in coastal regions but have been known to occur in inland lakes due to the passage of atmospheric pressure or shock waves. They can be generated by a number of mechanisms including seismic activity, volcanic eruptions, undersea and subaerial landslides and atmospheric pressure jumps. Their imprint on the coastal landscape can be similar to that left by extreme storm waves and as a result differentiating their signature can at times be difficult. It is often the case that where evidence, such as sand layers within otherwise fine-grained sedimentary environments, occurs near tsunami-generating sources these forms of palaeoevidence will be attributed to this extreme event. Where the evidence occurs in locations distant from tsunami-generating sources, particularly earthquakes, the evidence is often attributed to storm waves. Neither assumption, however, is strictly based upon sound logic because of the variety of mechanisms that generate tsunamis. However, in regions that regularly experience seismic activity and have a history of tsunami inundations such a conclusion seems reasonable.

Palaeotsunami records often show that tsunamis larger than those occurring historically have occurred in the relatively recent past. Classic examples are the records from the northwestern USA, New Zealand, the Mediterranean Sea and the North Sea. Australia too appears to have experienced larger tsunamis in prehistoric times although here the evidence is not so clearly distinguished from that of storm waves.

6

Earthquakes

Earthquakes and plate tectonics

The Earth's lithosphere or crust is broken into rigid plates that move away from, past and into each other (Fig. 6.1). This movement is not usually continuous but occurs in pulses. During such phases, the energy released is expressed as an earthquake. The majority of earthquakes worldwide occur at the boundaries between tectonic plates. As Abbott (1999) notes, there are three separate processes related to plate tectonic movements that can produce earthquakes. These are as follows.

(1) The pull-apart motion at spreading centres causing rocks to fail in tension. Rocks rupture sooner when subjected to tension. This process yields mainly smaller earthquakes that do not usually pose an especially great threat to humans.

(2) The slide-past motion that occurs as rigid plates wrap around the curved Earth. The plates slide past each other via dominantly horizontal movements of transform faults (transform faults are the fractures generated by rigid plates moving over Earth's curved surface). These movements can create large earthquakes because the irregular plate boundary retards slip along irregularly shaped faults. Considerable stored energy must be expended to overcome the rough surfaces of rocks and the bends in the faults. A large amount of seismic energy is released in overcoming these impediments.

(3) The collisional motions that occur at subduction zones and in continent–continent collisions. These motions release immense amounts of energy resulting in Earth's largest tectonic earthquakes. An incredible

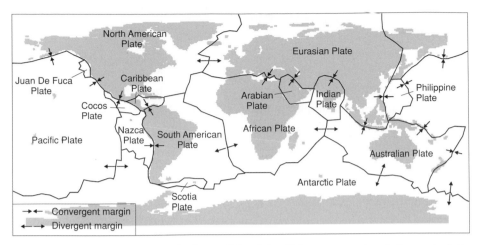

Figure 6.1. Major tectonic plates of the globe. Major convergent and divergent plate margins are marked by arrows. Most of the high-magnitude earthquakes occur along the convergent margins.

amount of energy is required to pull a 70–100 kilometre thick oceanic plate back into the mantle via a subduction zone. The same is true when continents push together, such as India pushing into Asia to create the Himalayas.

Earthquakes tend to be more severe along plate margins dominated by transform faults and subduction/collision zones compared to spreading centres (where tectonic plates are moving away from each other). The major tectonic plate slide past motions on Earth occur along four long transform faults. These are the Queen Charlotte fault in the Northeast Pacific, the San Andreas Fault in California, the Dead Sea fault zone in the Middle East and the Alpine fault at the southwestern edge of the Pacific Ocean which cuts across the South Island of New Zealand. The major global spreading centres are found in Iceland, the Red Sea, the Gulf of Aden, the Galapagos and Azores Islands, and the East African Rift Valley system.

Tectonic plates collide with each other in one of three main ways: ocean plates colliding with ocean plates, ocean plates colliding with continental plates or continental plates colliding with continental plates. The first two result in subduction of one plate beneath the other while the latter results in continental upheaval. Oceanic plates are normally subducted due to their greater density. Continental plates are usually not subducted as the huge volumes of low-density, high-buoyancy rocks cannot sink to great depth and be pulled into the mantle (although the Australian continental plate is being subducted below the Asian continental plate). Examples of continental plate collision include the Arabian plate pushing into Eurasia, and India colliding with Asia to create the Himalayas and the Tibetan plateau.

Three-quarters of the earthquake energy released on Earth occurs along these convergent plate margins. The majority of these earthquakes are shallow focus events that occur in the upper 60 km of the crust. Some earthquakes occur up to 700 km under the Earth's surface. Below this, little or no seismic activity is found.

Earthquakes also occur away from plate boundaries as crustal stresses are released due to continental uplift, crustal readjustment resulting from glacial deloading (isostasy), and strain release of energy by fracturing where the source of stress is likely to be thermal imbalances in the upper mantle.

Earthquakes and faulting can also occur due to dilatancy in crustal rocks. The pressure of overlying rocks equals the strength of the unfractured rock at depths greater than 5 km. Here, sudden brittle failure and frictional slip do not occur as the necessary shearing forces cannot be obtained. This is the case where rocks deform plastically. However, sudden rupture can occur if the effective friction along any crack boundary is reduced as a result of the presence of water. If rocks in the upper crust strain without undergoing plastic deformation, they can crack locally and expand in volume. This process is called dilation. Although dilation can occur too quickly for immediate groundwater penetration, water will eventually penetrate the cracks and provide lubrication for the remaining stress to be released.

Four different wave types are transmitted during an earthquake: P-waves, S-waves, L-waves (I) (Love wave) and L-waves (II) (Rayleigh wave). The P-wave is the primary wave and can be described as a compressional wave that spreads out from the epicentre. It consists of alternating phases of compression and dilation and can travel through gases, liquids and solids, as well as refracting at fluid–solid boundaries. The velocity of a P-wave is dependent upon rock density and compressibility. Although these waves have the potential to travel through the core of the Earth, they are refracted at the core–mantle boundary into two shadow zones 3000 kilometres wide. P-waves are not detectable on opposite sides of the Earth due to this refraction.

The S-wave is a shear wave and travels 0.6 times slower than P-waves. The velocity of S-waves depends upon rock density and rigidity. Although S-waves can travel through the mantle they cannot travel through the rigid core of the Earth. The shadow zone this creates on the opposite side of the core overlaps the shadow zones created by the P-waves. Because of this, the spatial distribution and time separation between the arrival of P- and S-waves at three separate seismograph stations can be used to determine the location and the magnitude of an earthquake.

L-waves are long surface waves that are trapped between the surface of the Earth and the crustal layers further down. These waves are not transmitted through either the mantle or the core, but spread relatively slowly outwards from the epicentre along the surface of the Earth at a speed of 5000 km in 20 minutes. The energy of these waves dissipates progressively with distance from the epicentre.

Earthquake magnitude and intensity

Earthquakes, along with volcanoes, release the most energy in the shortest time of all the natural hazards. Magnitude and intensity of earthquakes are measured using the Richter and Mercalli scales, respectively (Robinson, 1993). The Richter scale measures the total amount of elastic energy released by each shock wave during an earthquake. With this scale, magnitude (M) is a function of energy (E) and is described using the following equation,

$$\log_{10} E = a + bM \tag{6.1}$$

The values of a and b have been modified several times. In 1991 their values were approximately 5.8 and 2.4 (Bryant, 2005). These values define the released energy in ergs. One unit increase in energy on the Richter scale describes 2.4 orders of magnitude change, which can also be described as a 240 times increase in energy. Equation (6.1) only describes magnitude as a function of energy. It is not strictly applicable to changes in the amplitude of the shock wave although wave amplitude is often taken as a guide to the amount of energy. Each unit increase in magnitude on the Richter scale describes one order of magnitude increase in wave amplitude. One order of magnitude can also be described as a tenfold increase in wave amplitude.

Different estimates of the magnitude of an earthquake can be recorded on the Richter scale because of the different pathways that seismic waves take travelling to recording stations, differences in seismograph equipment and variations that still exist in the parameters used to describe magnitude by equation (6.1). The Moment scale was developed, in conjunction with the Richter scale, to help overcome these difficulties. It measures the exact energy released by an earthquake and is a function of the surface expression of faulting associated with the earthquake. At times, use of the Moment scale is limited because often surface expression only occurs in earthquakes of magnitude 7.0 or greater on the Richter scale (Bryant, 2005).

Table 6.1 *Modified Mercalli scale*

Scale	Intensity	Description of effect	mm s^{-2*}	On R. scale
I	Instrumental	Detected only on seismographs	<10	
II	Feeble	Some people feel it	<25	
III	Slight	Felt by people resting: like a large truck rumbling by	<50	<4.2
IV	Moderate	Felt by people walking; loose objects rattle on shelves	<100	
V	Slightly strong	Sleepers awake; church bells ring	<250	<4.8
VI	Strong	Trees sway; suspended objects swing; objects fall off shelves	<500	<5.4
VII	Very strong	Mild alarm; walls crack; plaster falls	<1000	<6.1
VIII	Destructive	Moving cars uncontrollable; chimneys fall and masonry fractures; poorly constructed buildings fall	<2500	
IX	Ruinous	Some houses collapse; ground cracks; pipes break open	<5000	<6.9
X	Disastrous	Ground cracks profoundly; many buildings destroyed; liquefaction and landslides widespread	<7500	<7.3
XI	Very disastrous	Most buildings and bridges collapse; roads, railways, pipes and cables destroyed; general triggering of other hazards	<9800	<8.1
XII	Catastrophic	Total destruction; trees driven from ground; ground rises and falls in waves	<9800	<8.1

*maximum acceleration measured in mm s^{-2} (from Bryant, 2005).

The modified Mercalli scale measures earthquake intensity. It is principally a qualitative scale that describes the type of damage occurring close to the earthquake epicentre. These damage descriptions are presented in Table 6.1. The scale categories increase with increasing acceleration of the shock wave through the crust. The interval between each category is approximately proportional to $\log_2$ until the maximum acceleration exceeds 9800 mm s^{-2}. This is the acceleration due to gravity and when this speed is exceeded by the acceleration of the shock

Table 6.2 *Large historical earthquakes as measured by magnitudes*

Magnitude	Location	Year
9.4	Alaska	1964
9.3	Banda Aceh	2004
9.2	Chile	1960
9.0	Lisbon, Portugal	1755
8.9	Sumba, Indonesia	1977
8.3–8.9	Chile	1960
8.7	West Sumatra (Niasisland)	2005
8.6	Andes, Columbia	1960
8.6	North Assam, India	1950
8.6	Alaska	1964
8.5	Kansu, China	1920
8.5	Japanese Trench	1933
8.4	Valparaiso, Chile	1906
8.4	Tienshan, China	1911
8.25	San Francisco, USA	1906
8.2	Tokyo, Japan	1923

wave through the crust of the Earth, objects can be tossed in the air and trees physically uplifted from the ground (Bryant, 2005).

High-magnitude historical earthquakes

Many large earthquakes have occurred in historical time (see Table 6.2). These have all exceeded 8 on the Richter scale and have often caused great hardship and considerable loss of life. One of the largest ever earthquakes experienced in modern times was the AD 1755 Lisbon earthquake. This event caused massive destruction in Portugal and also resulted in another earthquake 550 km away in Morocco. Ninety percent of buildings in Lisbon were destroyed or severely damaged and almost 70 000 lives lost. Structures as far as 600 km away suffered damage when tsunamis struck the coasts of Portugal, North Africa, the British Isles, the Netherlands and even the West Indies (Abbott, 1999).

The loss of human lives and damage to buildings is not always a reflection of the magnitude of an earthquake. Human impacts are often more a function of community vulnerability, such as the ability of buildings to withstand the earthquake waves. The saying '*earthquakes do not kill people, buildings do*' was well exemplified by the December 2003 Iranian earthquake where the human death

Table 6.3 *Deaths from earthquakes*

Death toll	Location	Year
830 000	Shensi, China	1556
300 000	Calcutta, India	1737
250 000	Tangshan, China	1976
200 000	Kansu, China	1920
143 000	Tokyo, Japan	1923
100 000	Chihli, China	1290
80 000	Caucasia, Shemaka	1667
75 000	Messina, Italy	1908
70 000	Northern Peru	1960
60 000	Cilicia, Asia Minor	1268
50 000	Bam, Iran	2003
45 000	Corinth, Greece	856
25 000	Armenia, Soviet Union	1988
700	San Francisco, USA	1906

toll exceeded 50 000. This earthquake, which struck the city of Bam, measured only 6.3 on the Richter scale. Despite its relatively moderate magnitude, the 2003 Iran earthquake ranks as the 11th worst earthquake in terms of human impacts (loss of life) over the last 1100 years (Table 6.3).

Building collapse and fire are two of the main causes of death during and in the aftermath of earthquakes. Buildings are often designed to withstand earthquakes in higher income nations where the earthquake hazard is high but this is often not the case in poorer countries and those that usually experience low seismic activity. It is often in these latter locations that earthquakes result in the greatest loss of human lives. One of the worst earthquakes in recent times, in terms of loss of human lives, occurred in Tangshan in China in 1976. Although an earthquake 18 months earlier had been successfully predicted and the city evacuated, there were no known clear signs of this impending earthquake. A quarter of a million people were killed due to building collapse and the fires that followed (Robinson, 1993).

Artificial dams have been known to cause earthquakes. One of the most severe earthquakes following the construction of a dam took place in Koyna in India in 1967. This is an area of generally low seismicity but tremors became a frequent occurrence after 1962 when dam construction began. In 1967, a number of sizeable earthquakes occurred prior to an intensity X earthquake on the modified Mercalli scale. The earthquake damaged buildings and killed more than 200 people (Robinson, 1993).

Other hazards associated with earthquakes

There are three main natural impacts resulting from earthquakes: liquefaction, landslides and tsunami. Liquefaction occurs where the pore pressure of water equals or exceeds the weight of the overlying soil. Soils behave as a solid when the pore pressure (pressure in voids or spaces between individual particles or grains) is lower than the weight of the soil (Bryant, 2005). But pore pressure increases substantially during earthquakes resulting in a loss of grain to grain contact within the soil. The increased pore water pressure suspends the particles creating a dense slurry and the soil behaves as a liquid. The liquefaction potential of a soil is dependent on the grain size of the sediments. Greater cohesion between silt and clay particles prevents pore water pressure from having the potential to equal the weight of the soil. Liquefaction, therefore, usually occurs in medium- to fine-grained sands typical of alluvial and marine sediments (Bryant, 2005).

Earthquakes cause landslides by weakening the structural integrity of natural slopes. Rock fall is the most common form of slope failure but snow avalanches have also been known to occur. A mainly rock and snow avalanche from a cliff overhang in the Nevados Huascaran mountains (Peru) following a magnitude 7.7 earthquake caused more than 50 million m^3 of mud and boulders to travel at more than 250 km h^{-1} as a 30 m high surge. More than 18 000 people died when the avalanche buried a number of towns (Chapman, 1999).

Earthquakes greater than magnitude 6.5 on the Richter scale have the potential to produce tsunamis (see Chapter 5). Earthquakes that produce tsunami are often shallow earthquakes that occur at depths between 0 and 40 km (Bryant, 2005). Earthquake-generated tsunamis are usually associated with subsea fault movements causing a vertical offset of the seabed. The peak energy period of a tsunami is a function of the magnitude of the earthquake and can be described using the following equation,

$$T_t = 0.57M - 2.85 \qquad\qquad (6.2)$$

T_t describes the period of the tsunami while M describes the magnitude of the earthquake on the Richter scale. Although the waves are relatively low at sea (generally <1 m), they shoal and reach extreme heights in shallow water close to the coast. Because shore lines are often relatively steep compared to the wavelength of the tsunami, the wave does not break but surges up over the foreshore. The run-up height of the wave also depends on the configuration of the shore, diffraction of the approaching wave, and the characteristics of the wave and resonance (Table 6.4). Tsunamis travel at speeds of up to 800 km h^{-1} in the deepest parts of the ocean and 300 km h^{-1} across the continental shelf.

Table 6.4 *Earthquake-generated tsunamis*
(run-up heights and deaths)

Year	Site	Height	Deaths
2004	Banda Aceh	+30 m	100 000
1896	Japan	29 m	27 000
1992	Indonesia	26 m	1000
1933	Japan	20 m	3000
1946	Alaska	15 m	175
1993	Japan	11 m	175
1755	Portugal	10 m	30 000
1960	Chile	10 m	+1250
1992	Nicaragua	10 m	+150
1964	Alaska	6 m	125

From Abbott (1999).

Tsunamis also resonate in harbours when the wave period of the tsunami is a harmonic of the natural frequencies of that harbour or bay. This can result in the amplitude of the tsunami increasing greatly over time (Bryant, 2005).

Earthquake prediction

Earthquake prediction is not an exact science. However, progress is being made in short-term predictions of earthquake activity. Approximately 3 months after the Boxing Day (2004) earthquake and tsunami in Sumatra another large earthquake (magnitude 8.7) occurred 200 km south near the Island of Nias, Sumatra. This earthquake was predicted only 2 weeks before it occurred by McCloskey *et al.* (2005). They calculated the stress imparted by the 2004 Boxing Day earthquake onto nearby structures and faults and identified two main zones that could experience an earthquake. One was the onshore Sumatra fault that runs through the Island of Sumatra. The other at risk zone was the offshore segment of the subduction zone and it was here where the March 2005 earthquake occurred.

Other studies have also presented methods to make short-term predictions of earthquake activity. McGuire *et al.* (2005) compared foreshock to aftershock activity on East Pacific rise transform faults. Through a retrospective study of past earthquakes they developed a model which can be used to monitor foreshock activity and predict approximately when and where a major earthquake may occur. There have been a number of reports of unusual animal behaviour in the days prior to earthquakes. It is not known why animals seem able to detect imminent earthquakes, but there appears to be little doubt that they are capable of this phenomenon (Robinson, 1993). Catfish for example were reported to jump agitatedly and could easily be caught immediately prior to the 1923 Tokyo earthquake. Panicked behaviour in rats is an official indication of an impending earthquake in China. Bryant (2005) provides more detailed descriptions of this interesting phenomenon.

PALAEOEARTHQUAKES

Earthquakes cause substantial, localised, changes to the form of the Earth's surface and near surface. This occurs as:

- sediment deformation (warping, microfaulting);
- deposition of various sedimentary units;
- formation and destruction of landforms; and
- deformation and destruction of human built features.

These events can also be recorded in tree rings and speleothems. Analysis of Quaternary sediments and landforms is the most common form of reconstructing ancient earthquakes.

There are several ways in which an earthquake may affect sedimentation and ground properties. These include:

- rockslides, landslides, mudslides;
- deposits falling from steep slopes and distant rolling of blocks and debris along slopes;
- formation of scarps, flexures, ravines and cracks;
- displacement of sedimentary and rock layers and bodies vertically and horizontally;
- warping and non-uniform settling or shaking in soils;
- cracking of hard or loose sediments;
- ejection of loose detrital materials forming local hollows and heaps;
- water saturation and liquefaction flow in soil, mostly clayey and sandy ones; and
- formation of channels and cones of sand eruption (sand blows).

Not all of these features are necessarily unequivocal evidence of a past earthquake for other events such as extreme meteorological and climatic changes, as well as human activities, can result in similar sediment and landscape features. When these other factors can be eliminated as a potential cause, these features are useful in locating areas affected by palaeoseismicity.

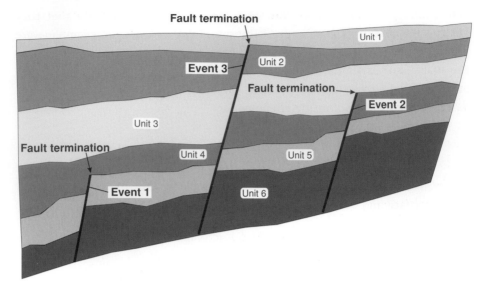

Figure 6.2. Idealised sequence of upward fault terminations in sedimentary units. Note the increased disruption of sedimentary units with depth. Three palaeoearthquakes are evident in this stratigraphy.

MICROFAULTS (UPWARD FAULT TERMINATIONS)

Microfaults or upward fault terminations are small faults that occur within sedimentary and rock units during major earthquakes. These faults are usually examined in late Quaternary sedimentary sequences that straddle major faults (Nikonov, 1995). Often these sedimentary sequences experience ongoing sedimentation with time, so early earthquake events will produce microfaults in the sediments that are present at the time. Subsequently deposited sedimentary units will not record the early earthquake because they were not present during that event. But these later sedimentary units can be faulted by subsequent major earthquakes. In this fashion, the microfaults terminate upwards against a sedimentary unit that was deposited after that earthquake event. The number of overlying sedimentary units displaying upward terminating faults provides information on the number of prehistoric earthquakes at that site and the ages of the lowermost faulted unit and overlying non-faulted unit bracket the timing of the earthquake.

Figure 6.2 illustrates a hypothetical microfaulting scenario. Unit 6, the oldest sedimentary unit is overlain by unit 5 and both are faulted. The fault does not extend into unit 4 so this unit provides a minimum age for the event while unit 5 provides a maximum age. In other words the earthquake occurred during deposition of unit 5. Unit 4 is disrupted by another fault that terminates at unit 3.

Hence the age of this event is bounded by the ages of units 4 and 3. Likewise, a third earthquake event is evident in the stratigraphy and it terminates at unit 1. There have been no earthquakes since deposition of unit 1 began, but when the next earthquake occurs it too will experience surface disruption. The older the unit the more earthquakes it will have experienced and the more deformed it will be compared to younger units.

When reconstructing these event stratigraphies it is important to determine that the downward increases in deformation were derived from sudden, brittle surface rupturing processes associated with large earthquakes rather than from gradual fault movements (Lienkaemper *et al.*, 1999). Fumal *et al.* (2002a) found that upward fault terminations may not always be a very reliable indicator of past earthquakes. They suggest that folded sedimentary units, where they can be mapped across a relatively broad area, may be more useful. Fumal *et al.* (2002a) used fault scarps, colluvial wedges, in-filled fissures and increased tilting of sediment layers downsection to identify earthquake horizons.

Studies using microfaulting and associated deformation features have shown detailed histories of major earthquakes along large fault systems in California, USA and elsewhere globally. The Hayward fault in northern California, which lies just east of the San Andreas fault, for example, has experienced seven or more surface faulting earthquakes over approximately the last 2000 years (Lienkaemper *et al.*, 1999). The average recurrence interval between these events is <270 years. The last major earthquake occurred here after AD 1640 but there have been no large earthquakes since AD 1776. This suggests that the next large earthquake in this region is relatively imminent.

A similar conclusion was drawn by Fumal *et al.* (2002b) for segments of the San Andreas fault, California. Four and probably five large earthquakes occurred on the Mission Creek strand of this major fault over the past 1200 years. The most recent large earthquake occurred around AD 1675 and the palaeoseismic record suggests that these events have a mean recurrence interval of around 215 years. This suggests that over 300 years have elapsed since the last major earthquake and the region is overdue for the next event.

Liquefaction features

Liquefaction features, or sand blows, occur when ground shaking during an earthquake liquefies a subsurface layer of sand which penetrates into and/or through an overlying, otherwise impenetrable layer. The overlying layer is intruded via a sand dike and if this layer is the uppermost stratigraphic unit the liquefied sand can penetrate to the ground surface and be expressed as a mound or area of sand. The original ground surface is commonly depressed as a result of the displacement of the subsurface sand and water.

Liquefaction of the sand layer takes place when shear waves propagate through saturated granular layers. These waves cause collapse of the granular structure, and in so doing can cause a significant increase in the intergranular pore pressure if drainage is impeded (Li *et al.*, 1996). When the pore pressure equals the weight of the impenetrable overlying sediment, the granular layer liquefies. This means that the sand behaves like a viscous liquid rather than a solid. The liquefied sand and water can then vent through fractures to the surface (Li *et al.*, 1996). Sims and Garvin (1996) propose a model to describe the formation of sand blows. First, as a function of the strong ground shaking, grain-to-grain disruption and suspension occurs. This is then followed by a pore-water pressure increase, a redeposition of the suspended sediment, fissure development and finally eruption of sediment and water at the ground surface. Matter from below the ground surface can often be found in both the sand blows and in the dikes themselves, and this material can show the direction of the flow of the vented sand (Li *et al.*, 1996). When the liquefied sand does not reach the ground surface, a liquefied zone at depth can be identified when that layer exhibits little or no shear strength and increased pore-water pressure. Critical to the liquefaction process is the increase in pore-water pressure. Measurements of *in situ* pore-water pressure during modern earthquakes, such as the 1987 Superstition Hills (USA) earthquake, show that the pore-water pressure increases gradually to equal the pressure of the overlying sediment, but does not peak until cessation of the earthquake (Sims and Garvin, 1995).

Sand blows have been observed and documented in relation to earthquakes worldwide. On a regional basis, the areal extent and concentration of liquefaction-induced features are proportional to the energy released by the earthquake. The location of individual sand blows, however, is determined by the age and distribution of potential sources of sand, the presence of an impermeable layer above the liquefied sand, the thickness of the impermeable layer, and the morphology of contact between the source layers and the impermeable layers. The development of liquefaction features is dependent on several factors such as the shaking intensity of the earthquake and the presence of liquefiable sediments. The size of sand blows is assumed to be related to the intensity of shaking during the earthquake. Structures formed at higher intensity shaking tend to be larger than features deformed during lower intensity events. Their preservation is largely dependent on the environment in which they are found (Sims and Garvin 1995).

Sand blows can be used as indicators of prehistoric earthquakes in areas that are seismically active but lacking clear surficial faulting. However, distinguishing between historic and prehistoric sand blows can be difficult. Historic sand blows have weakly developed soils displaying thin A horizons with a single grain or massive soil profile structure. Prehistoric sand blows have more mature soils with thicker A horizons and an angular, blocky or friable, or more mature structure. Tuttle and Schweig (1996) determined that soil A horizons within sand blows in the New Madrid seismic zone in the central USA develop at an average rate of 0.4–0.5 mm yr^{-1}. Adams (1996) drew similar conclusions regarding sand blows in Canada. In cases where the sand blows are buried, the thickness of the palaeo-A horizon can be used as a guide to the length of time the sand blow was exposed at the ground surface.

Tuttle and Schweig's (1996) estimates of the rates of soil formation in sand blows prove useful for dating prehistoric seismic activities in other regions where liquefaction has occurred. However, it is always important to remember that soils can develop at different rates under different climatic regimes. The same is true of regional variations in other soil forming factors such as topography, parent material, and faunal and floral variations. Independent evidence of rates of soil formation need to be ascertained for any individual region before any reasonable estimates of the age of prehistoric sand blows, and hence past seismic activity, can be made.

Archaeological evidence can prove useful in dating sand blows. Tuttle and Schweig (1996) noted that relatively well-drained, prehistoric sand blows in the New Madrid seismic zone were used by Native Americans as home, storage and burial places. Chronological determinations of artefacts found in association with sand blows, provide a minimum age for the seismic activity. Radiocarbon and luminescence dating techniques (see Appendix A) can be used in such settings. Luminescence dating is particularly useful if the sand blow is sufficiently thick so that the buried sands have been excluded from exposure to sunlight and cosmogenic rays. Radiocarbon dating of organic inclusions within the sand blow from sedimentary layers overlying the liquefied sand is also possible but this only provides a maximum age for the seismic event as organic inclusions are likely to be older than the event itself.

It is difficult to date and distinguish sand blows in regions where earthquakes have frequently induced liquefaction events. It is important that the ages of sand blows are well constrained if these liquefaction features are to be correlated regionally to determine the source areas and magnitudes of the prehistoric seismic events. Events that have occurred over time periods of less than a couple of hundred years are difficult to distinguish using conventional radiocarbon and luminescence dating. This is because of the uncertainty margins of 50–100 years, or more, associated with such dates; however, accelerator mass spectrometry radiocarbon dating can help to reduce these uncertainty margins.

When uncertainty margins from dated sand blows across a region overlap, those events cannot technically be distinguished and the possibility that the same earthquake generated a suite of sand blows must be considered. The only way to determine whether the sand blows occurred independently of each other is to use dating techniques of sufficiently high resolution so as to obtain small uncertainty margins. Dating liquefaction events can also be made more difficult if sand-bearing water has intruded the overlying sediment and formed sand dikes and sills without venting to the ground surface. Where this is the case, the liquefaction feature can only be dated based on the age of the uppermost intruded stratigraphic unit (Tuttle and Schweig, 1996).

There are several other problems associated with using liquefaction features to date prehistoric earthquakes. One of the most problematic is differentiating between earthquake-generated structures and liquefaction features formed by other geological processes that mimic seismic liquefaction events (Sims and Garvin, 1995). Non-seismic sand boils form when water seeps beneath levees during floods; these boils can mimic sand blows developed during seismic events (Li et al., 1996). Misinterpretations, where flood-induced sand boils are mistaken for earthquake-induced sand blows, can lead to a hypothesis of shorter earthquake recurrence intervals and thus an exaggeration of the seismic hazard. This can also result in an exaggeration of the magnitude of prehistoric earthquakes as the spatial extent of these events can be overestimated.

Li et al. (1996) use six criteria to distinguish between earthquake-induced liquefaction and flood-induced sand boils. These are:

(1) earthquake-induced liquefaction deposits are broadly distributed along an epicentral area, whereas flood-induced sand blows are limited to a small band along river levees;

(2) the conduits of most earthquake-induced sand blows are planar dikes, whereas the conduits of flood-induced sand blows are most commonly tubular;

(3) depression of the pre-earthquake ground is usual for sand blows, but not for sand boils;

(4) flood-induced sand boils tend to be composed of better sorted and much finer-grained sediments than sand blow deposits;

(5) source beds for earthquake-induced deposits occur at a wide range of depths, whereas the source bed for sand boils is always near the surface; and

(6) materials removed from the walls surrounding the vent of a sand blow are seen inside the sand blows. In general, flood-induced sand boils are interpreted to represent a less energetic genesis than earthquake-induced liquefaction.

While some authors have noted that the size and distribution of sand blows can be used as a guide to the size and distribution of prehistoric earthquakes, recent earthquakes have shown that this may not always be the case. The 1989 Loma Prieta, California and 1994 Northridge earthquakes resulted in an irregular distribution of liquefaction features that were not centred around the earthquake epicentre (Tuttle and Schweig, 1996). The irregular distribution of sand blows possibly occurred due to several factors such as the characteristics of the earthquakes themselves, the directivity, site conditions and Moho (boundary between Earth's crust and mantle) reflections of seismic waves. Since these factors are often unknown for prehistoric liquefaction features, it should be noted that there are large uncertainties associated with using the distribution of sand blows as a guide to the epicentre of a prehistoric earthquake. Nonetheless, where all other caveats have been checked, sand blows of some antiquity, either dated through an absolute or relative dating technique, can provide important information on the frequency of past earthquakes in a region.

Seismic deformation of muddy sediments

Fine-grained or muddy sediments that deform due to a range of factors are referred to as hydroplastic sediments. Hydroplasticity refers to the significant yield strength of a sediment–water mixture that is subject to cohesive or frictional forces. Although deformation structures of sandy sediments are well known and are characterised by liquefaction and water escape, the criteria used to determine deformation in muddy sediments are relatively poorly known and ambiguous. This is because hydroplastic sediments can be easily deformed by a number of factors including water flow or wave action on the sediment, the weight of overlying sediment, bioturbation and non-seismic slope failure. However, if these factors can be ruled out it is possible that the deformation occurred due to seismic activity. Matsuda (2000) notes that this is especially the case when the following occur:

(1) the site is near a seismic zone;
(2) there is a large lateral extent and synchronicity of deformation zones;
(3) liquefied and fluidised sediments are present;
(4) there exists a restricted deformation zone within a small vertical range of strata;
(5) limited horizontal displacement occurs;
(6) there are features that indicate *in situ* initiation of deformation and a cyclic deformation process;
(7) there are rare directional features in a structure; and
(8) there is a similarity to structures formed experimentally.

In most cases muddy sediment deformed structures are composed of sandy silts to silty clays and typically form in lacustrine sediments and settings. In a study of these features on the Kawachi lowland plain, Osaka, Japan, Matsuda (2000) found a homogenised layer in the uppermost part of the deformation zone, a plumose pattern in the middle to lower part of the zone, load structures below the plumose pattern, and downward fissures and microfaults in the lowest part of the zone. It is likely that these deformation features form when, during an earthquake, the sediment–water mix beneath a body of water (such as occurs in lakes) behaves rheologically from top to bottom. So the uppermost zone behaves as a liquid (liquidised), the zone below as a plastic (hydroplastic) and the lowermost zone as a brittle solid.

Matsuda (2000) found a vague mixing pattern in the sediments of the uppermost part of the deformation zone, especially towards the base of this zone. Sediments up to pebble size within the otherwise fine-grained sediments were probably derived via eddies from the underlying hydroplastic zone. The lower boundary of this zone was observed to take on a waveform in localised areas with a wavelength of 10–40 cm and an amplitude of 5–20 cm. There are two possibilities behind the formation of these waveforms. Either the wave movement of more cohesive units below forms them, or they are a result of the gravitational pressure of this unit and the shear force at the lower boundary layer. If the lower boundary is difficult to distinguish, it is assumed to have been caused by a gentle velocity gradient in the deformation flow near the boundary between the zones.

The plumose pattern found in the middle to lower section of the deformation zone (the hydroplastic deformation structure) forms when the sediments from the upper and lower zones mix (Fig. 6.3). Such features are recognisable because they display a different colour and texture from surrounding sediments. The plumose pattern consists of a series of slightly downward-convex flow lines and flow lines that trend upwards with frequent directional changes. These flow lines form where the sediment has been deformed in upward and downward directions. Sediments in this zone experience repeated scraping or dragging over the zone of lower sediments resulting in the latter being transported upwards. The distance between flow lines is usually less than 20 cm and the upper flow lines are slightly divergent or dispersed. Sometimes, in localised areas, the flow lines spiral upward in a counter-clockwise direction. This plumose pattern is, in most cases, easily distinguished and is assumed to form by instability at the boundary between the upper and lower flowing sediment layers. The mass transport responsible for creating the flow lines is assumed to occur by internal motion within an irregular standing wave. The poorly developed eddies and mixing of the sediments around the flow lines suggests that the mixing was

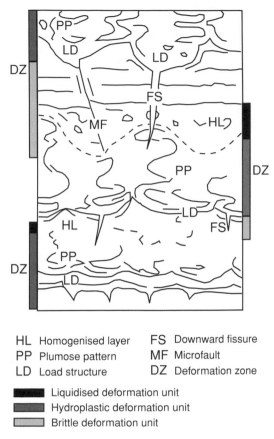

HL Homogenised layer   FS Downward fissure
PP Plumose pattern   MF Microfault
LD Load structure   DZ Deformation zone

■■■■ Liquidised deformation unit
▨▨▨▨ Hydroplastic deformation unit
▢▢▢▢ Brittle deformation unit

Figure 6.3. Schematic diagram of deformation structures in superposed deformation zones in muddy sediments (from Matsuda, 2000).

affected by the strong cohesive nature of the sediments; in this sense the zone containing plumose patterns can be regarded as a hydroplastic deformation unit. The morphological attributes of this zone are consistent with an intense deformation force that overcomes the shear strength of the cohesive sediments suggestive of deformation from seismic events.

Load structures may also be found at the lower reaches of the hydroplastic deformation layer (Fig. 6.3). Load structures are shallow downward-convex deformation features that penetrate into a lower stratigraphic unit or layer. They tend to occur at the base of the plumose patterns where the flow of the upper sediments is dragged over the lower sediments along a bedding plane. Their formation is assumed to be a result of an uneven downward pressure on bedding planes due to the oscillatory movement of the upper sediment during the earthquake event. These intrusions into the underlying sediment layer are not randomly distributed, but occur as circular patterns interconnected by the linear flow pattern. The shape of intrusions into the lower layer varies from circular to elongate arcs, and from round to angular. Most of the load structures also show a rotational flow pattern that is distributed with a preferred local orientation. This can be explained if the sediments of this unit migrated about 10 cm in one particular direction during the event. In some cases the hydroplastic load structures have been found in the next lower horizon that usually contains the brittle deformation structures. This suggests that the rheological response of the same sediment has changed during the seismic event.

Downward fissures and microfaults define the brittle deformation unit (Fig. 6.3). A downward fissure is a fracture that is developed vertically or obliquely and filled with the overlying sediment. The fractures frequently cross each other or bifurcate. They normally occur directly below the load structures. If the microfaults are densely distributed (several per 10 cm), they tend to show horizontal orientation, but adjacent microfaults tend to differ in orientation. Microfaults are assumed to form by a differential distribution of both vertical and horizontal forces.

There still remain uncertainties in using deformation features in muddy sediments as evidence of prehistoric earthquakes. This is principally because too few studies have been undertaken on this phenomenon. Their distribution globally, with respect to the major earthquake zones, along with insufficient knowledge of other processes that could possibly cause such features remains to be investigated. Their preservation in the recent geological record is also not well understood. It is also likely that they may not occur in certain environments and conditions. For example earthquakes that have occurred during the winter months in permafrost areas of Canada have apparently not left muddy deformation features. In such situations any palaeoseismic reconstructions in these areas would underestimate the frequency of such events. Past environmental and climatic changes in a region would also need to be considered when undertaking these studies.

Landform development (raised shorelines)

Coasts bordering subduction zones may rise or fall instantaneously during an earthquake (coseismic uplift or subsidence) and/or more slowly during interseismic periods (aseismic uplift or subsidence). Earthquakes with magnitudes above 8 that occur at the interface between subducting and overriding plates generally cause uplift of the region closest to the subduction trench. This process may also result in subsidence of the Earth's crust in a zone that is parallel to the uplifted zone. Vertical earth movements shortly before or after an earthquake may be as small or large as those during coseismic uplift. Japanese and Alaskan experience suggests the rates of aseismic movement generally decrease with distance from the subduction zone and with time following an earthquake. The long-term aseismic uplifts along coasts bordering subduction zones, however, occur at much slower rates than the short-term uplift occurring shortly after an earthquake.

Emergent shorelines are former shorelines that have been uplifted and no longer lie at an elevation where normal coastal processes can operate. Many such shorelines have arrived at their present elevation due to coseismic uplift during the Holocene. Historically, shorelines have been observed to experience rapid uplift during large earthquakes. As a consequence, it has been suggested that higher, older shorelines were likewise raised coseismically during seismic events. Where a number of raised shorelines are present, it is possible to estimate the recurrence and magnitude of the earthquakes responsible for their relative elevations. In this sense, the heights of these shorelines are taken to be indirect measures of the extent of plate-slip motion in the subduction zone. However, three major assumptions are required when these features are used to estimate palaeoseismic frequency and magnitude. These assumptions are:

(1) that long-term uplift rates are constant;
(2) the age and height data should approximate either time-predictable or displacement-predictable models of earthquake recurrence; and
(3) each great earthquake should be represented by a separate shoreline that is separate from shorelines that have formed during aseismic processes.

The third point is the most important of the three assumptions because it is often difficult to distinguish coseismic and aseismically uplifted shorelines along coasts bordering active tectonic margins. This is because storms can build successively lower beach berms on gently sloping coastlines that are rebounding from glacioisostatic depression. Also, the rates of sediment supply may exceed the rate of any sea-level rise – hence the coast appears to be experiencing uplift but is in fact accreting. It is also possible for large storms to reoccupy emerged strandlines (shorelines) and rework previously deposited beach sediments.

Reconstructing palaeoearthquakes using this method first requires identification of individual shorelines and then determinations of their elevation and distribution. Uplifted shorelines are often marked by sandy beaches or marine terraces cut in consolidated bedrock, including coralline limestones. The elevation of the shorelines can be measured from aerial photographs and in the field using standard topographic surveying techniques. Lithological and stratigraphic analysis of identified beach deposits (or former coral reefs) must also be undertaken. Most beach deposits are well sorted sands or gravel with faint planar bedding or laminations that strike parallel to the strandline. These sedimentary structures are typical of beach swash zones. Former shore platforms cut into bedrock are also good evidence of repeated coseismic uplift.

Shorelines can be dated using radiocarbon ages of molluscan shells and corals in both beach and alluvial deposits. The approximate maximum age of shorelines and their elevation can be used to calculate the sea-level history for a region in a manner similar to that undertaken for the Huon Peninsula in Papua New Guinea and in Barbados (Chappell and Polach, 1991).

Nelson and Manley (1992) identified as many as 18 raised shorelines up to 34 m above modern sea level in southern central Chile. These shorelines have been uplifted over the past 6000 years. It is uncertain how many of the shorelines were uplifted due to coseismic events (palaeoearthquakes) and for this reason it was difficult to calculate average recurrence intervals and average amounts of uplift per event. Most problematic were accurate identification, correlation and accurate dating of the older shorelines because of their close elevation spacing, subdued beach morphology, the lack of exposed wave cut platforms and uncertainties in the radiocarbon dating. Nelson and Manley (1992) noted that there was considerable uncertainty in the marine carbon reservoir effect (carbon mixes more slowly in the oceans than the atmosphere and this is corrected for in radiocarbon dating of marine shells) for nearshore marine shells at this latitude. This is mainly due to a lack of other studies along the same latitude. The large error intervals on the calibrated radiocarbon ages from these shorelines only made it possible to calculate average rates of sea level change over periods greater than a few hundred years. The large standard deviations of the ages also created large overlaps in the dating of the shorelines; some dates that are statistically similar were collected from shorelines separated by as much as 12 m elevation. Some shorelines at the same elevation also returned ages that differed by as much as 4100 years. Such differences can arise from the movement of the shells deposited on the shorelines by subsequent extreme wave events. Also, older shells can be incorporated into a shoreline and hence skew the age distribution of shells.

While these problems were not as serious for the younger shorelines, it is clear that a more detailed radiocarbon chronology is required if a meaningful magnitude and frequency record of past great earthquakes is to be obtained from this site. Nelson and Manley (1992) also thought it necessary to conduct a more thorough stratigraphic analysis to differentiate individual shorelines.

Stiros et al. (2000) identified three former shorelines, now uplifted, on Samos Island in the Aegean Sea. As well as the raised shorelines, microbenches and wave-cut notches are evident along the northwest coast of the island. The raised shorelines are 0.6, 1.1 and 2.3 m above sea level. Stiros et al. (2000) estimate that the shorelines were uplifted during earthquakes that occurred approximately 500, 1500 and 3600–3900 years BP.

Point measurements of surface rupture

Earthquakes usually occur due to fault movements and often the magnitude of an earthquake is proportional to the extent of that fault movement. The larger the fault movement and, therefore, displacement between the rock units separated by that fault, the more intense the earthquake. Measurements of this displacement therefore can be used as a guide to the earthquake magnitude. Two types of displacement can be measured: the length of surface ruptures and the extent of surface displacement. Often only the lengths of surface ruptures are used for this purpose rather than measurements of surface displacements. Surface rupture lengths have been commonly used because:

(1) most data for historic ruptures include rupture lengths that can be compared and compiled to estimate magnitude;
(2) magnitude estimates are most reliably correlated to fault length in historic data sets;
(3) measurements of fault length are easily acquired from geologic maps, seismicity plots and air photos; and
(4) segmentation schemes can be used to break large faults into rupture segments.

Hemphill-Haley and Weldon (1999) note, however, that there are several reasons why surface rupture lengths may not be reliable. These are:

(1) rupture length estimates are based on identification of often subtle, fragile geomorphic features that are easily eroded or buried, especially along long-recurrence faults;
(2) recent earthquakes have demonstrated that surface ruptures can integrate faults that were previously not known to be related;
(3) there are difficulties in assessing single event rupture segments on even the most active, mature faults with relatively short recurrence intervals; and
(4) segmentation schemes are generally not quantifiable in the sense that one can assign an uncertainty to the choice of a segment boundary. Often this last factor is handled with decision trees, where scenarios are weighed, but this approach cannot result in true uncertainties because only a few possibilities are considered and weights are assessed by experience that is often shared by the experts who weigh the scenarios.

Because of these problems it is likely that coseismic surface displacement is a better parameter for estimating prehistoric earthquake magnitude than surface rupture length. Establishing the extent of surface displacements is far more accurate than attempting to establish the boundaries of surface rupture lengths as the latter are often affected by geological and climatic processes over time. The magnitude of the palaeoearthquake can be estimated by comparing the mean displacement and magnitude of modern earthquakes with the mean displacement of the surface rupture. This can be achieved by:

(1) determining the slip associated with modern earthquakes of known magnitude before any measurements of palaeoearthquake slip are undertaken; and

(2) calculating the mean displacement for the whole surface rupture and then determining the statistical uncertainties from the range of preserved displacements as any one site may not be representative of the displacement of the entire rupture.

An assumption is made here that the relationship between the distribution of measured displacements of palaeoearthquakes is similar to that occurring in modern earthquakes. There is little doubt, however, that differences are likely to occur between the two data sets due to erosion and burial, and also geological sampling differences between historic and prehistoric displacement sites. Estimates of palaeoearthquake magnitudes using surface displacements have so far not distinguished between different earthquake styles, for a further assumption is often made when using this method that earthquakes create similar geological evidence over both time and space. It is likely that this assumption is to a certain extent incorrect. Nevertheless, displacement curves for different earthquakes probably do not display very significant differences. Such differences are also likely to be minimised through the random sampling of a number of displacements of both prehistoric and historic displacements (Hemphill-Haley and Weldon, 1999).

Landslide dammed lakes

Most earthquake-triggered landslides are rock avalanches or rock falls. The rocks fall rapidly from ridge crests and form a pile at the base of the slope and sometimes as shallow features against the opposite wall of the valley. These landslides can result in the creation of lakes due to the damming of streams. Such features have been observed in New Zealand, the United States of America, central Asia and the Himalayas. Adams (1981) suggests that these dams are sufficiently common to be used to estimate palaeoseismicity. Two types of lakes can result from earthquake triggered landslides. These are temporary lakes formed after damming the main trunk stream in a river valley; these features are ephemeral as river floods soon remove them. The second type is a permanent lake that forms on tributaries; these can remain intact for thousands of years because the tributaries have comparatively lower discharges and hence erosive power. Two factors, therefore, determine the longevity of a landslide-generated dam; the size of the landslide and the size of the river dammed.

The chronology of landslide-generated lakes is often determined using radiocarbon dating. Three approaches can be used: radiocarbon dating of wood in the slide debris, dating of standing trees drowned by the lake and dating of submerged soil horizons cored below lake sediments. It is possible to estimate the cause and size of a palaeoearthquake if a sufficient number of landslide-dammed lakes can be investigated. A series of landslide-dammed lakes of a single age can be used as an indication of the area shaken in one earthquake event. If care is taken in identifying these landslides they can provide a conservative estimate of palaeoseismicity that does not rely on the identification of an active fault. However, dating only one dammed lake is insufficient to give any conclusive answers. To establish the cause and size of an earthquake it is necessary to identify and date many lakes within the affected area (Adams, 1981).

Lake sediments

Earthquakes can cause sediments, often from landslides or slope movements, to be deposited into existing lakes and fiords. Records of prehistoric earthquake activity can be obtained through the study of these sedimentary sequences. The sediments are often in the form of turbidites which occur when sediments slump either subaerially or subaqueously. Turbidites often show a distinctive stratigraphy known as a Bouma sequence (although the Bouma sequence is now often replaced by facies models that separate low-density from high-density turbidite flows) and can be easily recognised from other sedimentary sequences. Apart from turbidites, sediments in these water bodies can also reflect changes, typically increases, in sediment influx. These changes can result from earthquake-induced erosion of the basin sides as a result of seiching (where waves are generated) and liquefaction of sediments and soils on slopes, as well as landslides. It is important to note that similar rapid changes in sedimentation can also occur due to climatic changes, volcanic activity and non-seismic effects. As with other sedimentary palaeoseismic records, it is important to rule out these other potential causes before an assumption is made that the sedimentary evidence is due to prehistoric earthquake activity.

Samples of these sedimentary sequences are often taken in the form of cores obtained from aboard a vessel. The cores can be several metres long and are often sliced into smaller sections before being stored in refrigerated conditions. The cores are analysed via detailed visual description, X-radiography, magnetic susceptibility, major element analysis, grain size distribution and microfossil presence and abundance. Sedimentation rates can be determined after obtaining a series of radiocarbon ages at selected intervals, or from sections of the core that appear to show distinct changes in sedimentation style.

Karlin and Abella (1996) undertook a study of sediments in Lake Washington, USA, in order to investigate the history of large earthquakes in the northwest Pacific region. They used magnetic susceptibility profiles to characterise the sediment lithology, which they found to be more reliable than grain-size measurements alone. The magnetic susceptibility analysis was able to identify concentrations of magnetic minerals, which better identified individual sedimentation events. In most of the samples, a peak in the magnetic susceptibility was associated with a 1100 year old turbidite layer near the top of the cores. Karlin and Abella (1996) attributed this layer to a palaeoearthquake, which had also been identified in studies using tree-ring analysis (Jacoby et al. 1992). Subsidence of the land during this earthquake caused trees, formerly growing on the lake shore, to become submerged into the lake water. The magnetic susceptibility peaks also varied directly with concentrations of an aluminosilicate (Al_2O_3) and inversely with concentrations of silica (SiO_2) and organic matter content as measured by loss-on-ignition (LOI). These results suggest that the sediments deposited into the lake over the past 3000 years are a mixture of a terrigenous fraction of detrital aluminosilicates and a lake-derived biogenic fraction of silicates (diatoms) and organic matter. The silica and organic carbon reflect more normal lake sedimentation. The combination and alternating relative concentrations of these sediments and the aluminosilicates suggest that there had been periodic changes in the terrestrial detrital sediment influx which is probably a function of earthquake-generated sediment fluxes to the lake.

The presence and absence of shallow water planktonic and non-planktonic species within sediment cores from lakes in these settings can also help identify major changes in sedimentation or slumping of the basin walls in prehistoric times. Organisms that normally live in shallow water, and the narrow littoral edge of a lake, can be transported rapidly into deeper water sediments through sudden slumping of the basin walls. Karlin and Abella (1996) also observed such changes within their cores from the deeper water sediments in Lake Washington. A normally river-dwelling species was found in cores from the centre of the lake, at the same core interval where the aluminosilicate peak occurred, suggesting there had been an abrupt, short-lived change in the character of the water. There was also a notable increase in non-planktonic organisms at this level of the cores suggesting a major influx of shallow water sediments to the deeper waters of the lake at this time.

One of the potential problems associated with using lake sediments and turbidites as evidence for past earthquakes is the difficulty in estimating the carbon reservoir correction of those sediments, for radiocarbon determinations. Such corrections are important because old carbon can be contained within the sediments due to the carbon residence time in the catchment sediments before they were deposited and submerged in the lake. The analysis is made easier if there are calibration ages present, such as a known and accurately timed event. When trees are submerged due to land subsidence during the earthquake, high-precision dating of the outer rings of these trees can be used to calibrate the radiocarbon dates from the lake sediments.

Archaeological evidence of prehistoric earthquakes

Archaeological evidence can help refine the temporal boundaries of past seismic events. This evidence can be relatively easily dated either through absolute geological dating techniques or through an understanding of the chronological changes experienced by various cultures, e.g. Chinese dynasties, Greek and Roman architectural styles and occupation of various lands. The types of evidence used to date prehistoric earthquakes often involve damage to buildings and infrastructure of a known age, and/or the cessation of occupation of a town/city or region. In such situations the age of the buildings or features provides a maximum age for the earthquake. A minimum age for the event can be determined by the age of undamaged buildings or features at the same site. These undamaged features are assumed to have been constructed after the earthquake. The ages of the damaged, and presumably later constructed undamaged features, bracket the age of the earthquake. Of course bracketing an event in this manner requires humans to have continued to occupy the site after the earthquake.

Archaeological evidence can also assist palaeoseismological techniques where occupation horizons in buried sedimentary sequences are cut across by sand blows and/or microfaults. In such situations the palaeoseismological features must post-date the archaeological evidence. The most commonly found artefacts used for dating of historic cultural sequences, and occupation horizons, are pottery shards and charcoal. Dateable objects can also include plant remains from agricultural activities or those occurring naturally and associated with a particular climate, possibly conducive to human occupation of a region for a period of time. Objects and plant evidence can be matched with artefact types and cultivation patterns that have been dated on other similar sites. In many regions in the United States, changes in artefact styles and types, and plant cultivation, have been mapped for the last 12 000 years (Lafferty, 1996).

Greek, Hebrew, Assyrian, Roman, Byzantine and Islamic cultures repeatedly reported the occurrence of large earthquakes throughout the Middle East over the past several thousands of years. Arab chroniclers described with details main shocks, aftershocks, surface breaks and related damage distribution in the Middle East as early as the 7th and 8th Centuries. The same Arabic documents also show that earthquakes have occurred less frequently over the last 830 years, or since medieval times, along a section of the Dead Sea Fault known as the 70 km long Missyaf segment. Because of this inactivity, the northern section of the Dead Sea Fault in Lebanon and Syria is considered to be an inactive seismogenic structure compared to other worldwide major strike-slip faults. Such assumptions, however, can lead to an underestimation of the risk associated with earthquakes in such locations, particularly if similar seismic gaps have been common throughout the late Quaternary (Meghraoui et al., 2003).

Meghraoui et al. (2003) undertook both an archaeoseismic and palaeoseismic investigation of a site along the Missyaf segment of the fault in Syria. The archaeological evidence included a disrupted Roman aqueduct which spans the Dead Sea Fault. The fault has experienced several episodes of left-lateral displacement totalling 13.6 m during major earthquakes and this has resulted in severing and relative displacement of each side of the aqueduct (Fig. 6.4). The age of the aqueduct was bracketed to between AD 30 and 70 corresponding to the early Roman period in the Middle East (post 64 BC). The geological evidence for large past earthquakes resulting from movements of the Dead Sea Fault at the site include faulted young alluvial, colluvial and lacustrine deposits, deflected streams with consistent left-lateral displacements of tens to hundreds of metres, and evidence of large shutter ridges and small pull-apart basins along the strike.

Palaeoseismic analysis involved trenching sediments at the site. The sediments contain pottery shards and detrital fragments of charcoal that were radiocarbon dated. A 5 m deep section exposed seven separate units of alluvial sediments. A 2 m wide shear zone with intense deformational structures and numerous fault strands occurs in the sediments. This shear zone formed during a number of large palaeoearthquakes. Sediments comprising the uppermost unit (unit a in Fig. 6.5) are not sheared or faulted but rather cap these features. This unit therefore has been deposited after the most recent earthquake and provides a minimum age of AD 990–1210 for this event. The next stratigraphically lower unit (unit b Fig. 6.5) truncates two faults that have occurred after deposition of the fourth highest stratigraphic unit (unit d Fig. 6.5) which was radiocarbon dated at AD 680–890. This unit itself caps another sedimentary unit

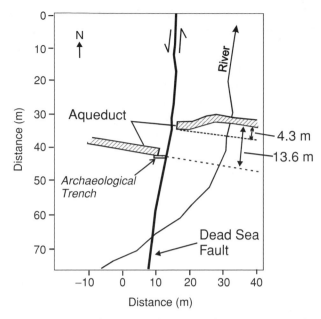

Figure 6.4. Map of the Missyaf segment of the Dead Sea Fault, Syria showing the left lateral displacement of an aqueduct (after Meghraoui *et al.*, 2003).

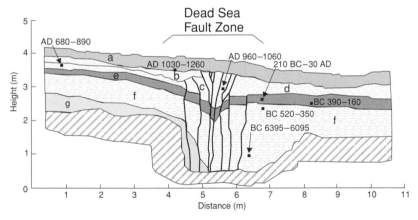

Figure 6.5. Stratigraphic cross-section of faulted sedimentary units along the Dead Sea Fault (after Meghraoui *et al.*, 2003).

(unit e Fig. 6.5) that was also affected by a large earthquake that occurred between AD 100 and 750. The palaeoseismic analysis therefore identified three large earthquakes occurring at this location over the last 2000 years.

Meghraoui *et al.* (2003) were able to confirm these palaeoseismic events with the Arabic historical record. The palaeoseismic record also allowed them to gain an appreciation of the magnitude of these events that they estimated between M_w7.0 and 7.5. The ages of these three events, and the extent to which the aqueduct spanning the fault had been offset or displaced, also allowed them to determine that the fault had experienced a late Holocene slip rate of 6.8–7.0 mm yr^{-1}. The slip rate and extent of displacement suggest an average recurrence interval for large earthquakes at this location of 550 years. The passing of 830 years since the last of these earthquakes highlights, firstly, that the fault zone is not inactive and, secondly, that the next large earthquake may be imminent, or possibly overdue. Such detailed conclusions would not have been possible from the otherwise excellent historical record.

The archaeo-palaeoseismic record of many seismically active regions of the world can be investigated in a manner like that undertaken for the Syrian section of the Dead Sea Fault where those lands have been occupied by ancient cultures. Iran, for example, has a long history of human settlement including a long-written record from the mid-8th Century. However, published palaeoseismic records in Iran are rare which limits the use of the archaeoseismic data. Berberian and Yeats (2001) note that some of these limitations include a lack of radiocarbon chronologies, the size of a destroyed area is poorly known for specific earthquakes, and it has been difficult to prove that an archaeological site was occupied continuously in the period between severe earthquakes. They do note, however, that estimating a maximum time interval between large-magnitude earthquakes can compensate for this latter uncertainty. Other problems have included the discovery of too few sites that provide evidence of damage from the same earthquake. This in itself prevents the mapping of meizoseismal areas and estimation of the moment magnitudes (Berberian and Yeats, 2001) of palaeoearthquakes. In the absence of palaeoseismic studies it can also be difficult to determine whether earthquakes or other events such as sudden abandonment of sites due to invasions, revolutions, climatic change or fire caused the damage to archaeological sites.

Studies in the New Madrid seismic zone, USA, have combined the dating of archaeological artefacts and liquefaction features to obtain a clearer seismic history in the area (Saucier 1991; Tuttle and Schweig, 1995). The sand blows here cross cut occupation horizons. Soil development in the liquefaction features is subtle in many cases and would be hard to date accurately without the presence of Native American cultural horizons and artefacts (Tuttle and Schweig, 1995).

Certain kinds or styles of damage to ancient buildings can also be indicative of past large earthquakes. The structural integrity of ancient buildings results from both static and dynamic effects. Static effects refer to building failure due to the weight of the construction itself without any external loading. Dynamic failures are caused by external violent loads such as earthquakes and other extreme events like wind, explosions, rock falls, floods and thunderstorms

(Stiros, 1996). Stiros (1996) has identified seven criteria for the recognition of earthquake-induced structures. These are:

(1) ancient constructions offset by seismic faults;

(2) skeletons of people killed and buried under the debris of fallen buildings;

(3) certain abrupt geomorphological changes, occasionally associated with destruction and/or abandonment of building sites;

(4) characteristic structural damage and failure of constructions, including:
  - displaced drums of dry masonry columns,
  - opened vertical joints and horizontally displaced parts of dry masonry walls,
  - diagonal cracks in rigid walls,
  - missing triangular sections of corners of masonry buildings,
  - cracks at the base or top of masonry columns and piers,
  - inclined or sub-vertical cracks in the upper parts of rigid arches, vaults or domes, or their partial collapse along those cracks,
  - keystones that have slid downwards in dry masonry arches and vaults,
  - several parallel fallen columns,
  - several fallen columns with their drums in a domino-style (imbricated) arrangement,
  - constructions deformed by horizontal forces (rectangles deformed to parallelograms);

(5) destruction and quick reconstruction of sites with the introduction of 'anti-seismic' building construction techniques, but with no change in their overall cultural character;

(6) well-dated destruction of buildings correlating with historical (including epigraphic) evidence of earthquakes; and

(7) damage or destruction of isolated buildings or whole sites for which an earthquake appears to be the only reasonable explanation.

It is important to be able to differentiate building damage due to earthquakes from that caused by human actions during battles and wars. The above criteria are designed to help in this differentiation and identify forms of evidence that would be difficult to induce by human actions alone. For example, horizontal forces or shaking are required to displace downwards a keystone in an arch. The same is true of columns whose drums show offset in different directions and particularly if those offset drums display a sinusoidal pattern along the length of column. Figure 6.6 shows such offset drums in a column from the Heraion Temple near Pythagorion on the island of Samos, Greece. The displaced columns

Figure 6.6. Offset drums in a column from Heraion Temple, Samos Island, Greece. The sinusoidal offset of drums was caused by horizontal forces during an earthquake after 530 BC. Photograph courtesy of Professor S. Stiros, Patras University, Greece.

show that the structure experienced horizontal forces from an earthquake that post-dated the construction of the temple around 530 BC (Stiros, 1996).

Tree-ring records (dendroseismology) and forest disturbance

Tree rings can record past earthquakes via a number of processes. One of these includes the survival of trees following seismically induced landslides. Surviving trees suffer damage such as toppling, tilting, trunk impact or root breakage and these impacts are recorded in the growth rings of the affected trees. The oldest undisturbed tree on a landslide provides a minimum age of the event. Along with landslides, the annual rings in disturbed trees can also be analysed to date recurrent landslide movements and to determine the magnitude and frequency of debris flows. Some of these methods have been used to investigate the possible relationship between landslide movements and regional seismicity in the Gravelly Range in southwestern Montana, USA (Carrera and O'Neill, 2002).

Trees can also show indirect responses to large earthquakes due to environmental changes. Sheppard and Jacoby (1989) noted that trees died from flooding caused by a general topographic reversal during the AD 1811–1812 New Madrid, USA earthquakes. Stands of undamaged trees experience increased growth when located near to areas of earthquake damaged trees because of the favourable light conditions after felling of the damaged trees. Trees also respond with increased growth in areas where coseismic uplifting has moved a shoreline away from a zone that was previously impacted on occasions by waves that inhibited growth of those trees. Their new position in the landscape affords them less exposure to wind, salt spray and root-zone erosion (Sheppard and Jacoby, 1989).

There are three common distinguishing features in tree rings that can be used to indicate past earthquakes. These are:

(1)   an abrupt reduction in annual ring width;
(2)   the formation of reaction wood; and
(3)   scars.

Disturbances found in the tree-ring record in a specific year indicate that the disturbance took place between the end of the previous growing season and the growing season of that specific year. A reduction in annual ring width for several years or more can be the result of injury due to a geomorphic event such as an earthquake-induced landslide or avalanche. Damage to the root system, loss of a major limb or toppling can also result in an abrupt reduction in annual ring width. When investigating the effects of the AD 1959 Hebgen Lake, USA earthquake on Douglas firs growing along the fault scarp, 13 of the 15 trees present exhibited a reduction in annual tree-ring width at the beginning of 1960. Although most of these trees showed signs of physical damage such as root breakage and tilting, the main response consisted of markedly narrower rings for 1–3 years after the event (Carrera and O'Neill, 2002).

Reaction wood is formed on the underside of a tree when a tree is tilted. The tilting results in asymmetric ring growth that is expressed as wide annual rings of reaction wood on the underside and narrow annual growth rings on the upper side of the tilted tree. Reaction wood in conifers is darker and denser than late wood and is usually characterised by a reddish-yellow colour and small, thick-walled cells. Although the formation of reaction wood generally dates the exact time of tilting, there is evidence that the formation of reaction wood can be delayed for several years after the event. This can occur where the tree is severely damaged and narrow annual growth rings form, as an initial response, on both the upper- and underside (Carrera and O'Neill, 2002).

Scars occur when the bark is removed and the underlying growth tissue (cambium tissue) is damaged. The tree will initially be unable to create annual growth rings in the wounded area. Over time, the growth tissue from the sides of the wound will overlap that wound and form a scar. At times, wounds on old trees can heal completely and as a consequence be difficult to detect (Carrera and O'Neill, 2002).

Sheppard and Jacoby (1989) note that it is important that only trees clearly affected by an event are used as event-response trees. It is best that trees are sampled based on age, size, and topographic and geologic setting. The proximity of tree responses to faulting varies with the nature of that movement. Dip-slip faulting results in wide disturbance zones whereas areas of disturbance near strike-slip faulting are restricted to within a few metres of the fault. Both disturbed trees and undisturbed trees must be sampled as the former provide a control for the latter. The use of controls is important in order to eliminate all other potential causal factors. Events such as wet or dry seasons can also affect the growth of annual rings. These effects can be distinguished from seismic events because seismic events in the tree-ring record display an extended period of recovery whereas responses due to climatic changes or influences are evident only for that particular growth season (Sheppard and Jacoby, 1989).

Multiple cores are normally taken from trees by non-destructive measures such as hand- or power-driven corers in order to help eliminate misinterpreting climatic or other non-seismic responses. Multiple cores are essential as non-uniformity of tree rings can occur in trees. Ring-width patterns are matched across different trees and all rings are assigned exact year dates. This use of cross dating accounts for any growth anomalies. After cross dating, the ring-width patterns are checked for disturbances. Disturbances other than the hypothesised seismic event can be identified by a growth response occurring synchronously in at least two separate trees. The ring-width analysis from undisturbed trees is combined into a site and species specific control series. Ring-width plots from disturbed trees are then compared to the control chronology in order to identify episodes of severe growth change resulting from past seismic events (Sheppard and Jacoby, 1989).

Large earthquakes can also induce significant forest disturbance where trees are literally shaken to the ground resulting in high mortality rates. Where trees are long lived, the age structure of these vegetation communities can reveal a story about the history of earthquakes when other possible disturbance mechanisms are eliminated. Wells *et al.* (2001) were able to identify four major earthquake events on the South Island of New Zealand, based upon the age structure of forests in the Westland region. These earthquake events occurred around AD 1820–1830, 1710–1720, 1610–1620 and 1460. They found that over 80% of the 1412 ha forested area studied currently comprises simple, first generation cohorts of trees established after the last catastrophic earthquake disturbance. This left only about 14% of the area inhabited by complex mature all-aged forest. This means that there appears to be a region wide over-abundance of mature trees and a lack of small to intermediate size trees. This feature of the demographics of the forest population had long intrigued scientists studying the area. This 'regeneration gap' had previously been thought to be possibly due to climatic cooling, an increased frequency and severity of droughts and decreased precipitation. However, Wells *et al.* (2001) suggest that major earthquakes play a significant role in structuring the forest ecosystems and that the reasons for the regeneration gap is likely to be due to the long period of time since the last major earthquake on the Alpine Fault which was about 280 years ago. It is likely that three of the major disturbances to the forest were due to earthquakes on the Alpine Fault and the last disturbance was due to an earthquake to the south of the region in AD 1826. Interestingly, it appears that this last earthquake caused a substantial tsunami around the shores of New Zealand's South Island which has been identified in sediment cores from Okarito Lagoon, Westland (Goff *et al.*, 2004).

Coral records of earthquakes

Coral microatolls can record earthquakes where they are subject to episodes of submergence or emergence due to vertical movements of the Earth's crust during major earthquakes. Coral microatolls are a reliable recorder of sea levels. The upward limit of coral growth is limited by the lowest tides above which the coral becomes exposed to the atmosphere and can die. Coral microatolls grow through deposition of annual layers of calcium carbonate. The shape of these layers reflects the nature of any sea-level changes occurring during the life of the coral (which can be several hundred years). During submergence of the coral the uppermost layer, which is normally restricted by the lowest tide level, experiences unrestricted growth and will form a 'cup shaped' morphology. This can contrast to the earlier deposited layers which may be flatter in shape reflecting a period of more stable sea level. During emergence a coral head devel-ops a conical shape because of the progressive fall in the level of the lowest tide. Episodes of sudden emergence and submergence are typically caused by vertical displacements of the Earth's crust during major earthquakes. During episodes of sudden submergence the microatoll layering shows a steep upward step from a flatter surface. During sudden emergence the microatoll develops a 'hat-shape' morphology where the hat 'brim' represents the continued growth for decades after the event. The step from top of the bowl to the brim is a measure of the vertical displacement during the event.

Natawidjaja (2002) analysed a 250 year long record of vertical deformation from four coral microatolls above the Sumatran subduction zone. The most prominent signal recorded by the corals is an 80 cm emergence event associated with the M_w7.7 earthquake in AD 1935 which resulted in a 2.3 m slip on the subjacent subduction interface. A 17 cm emergence occurred in AD 1797 and a greater than 10 cm emergence occurred in AD 1743. Natawidjaja (2002) estimates that the average slip on the subduction zone would have been about 0.5 m ($\sim M_w$7.0) in 1797 and at least 0.3 m ($\sim M_w$7.0) in 1743.

Earthquakes in AD 1833 and 1861 also appear to be registered by the 250 year long coral record. These events are the smallest earthquakes recorded here resulting in a 10 cm die-down of the coral head growth in 1833 implying about 30 cm of slip on the subduction interface, and a 5 cm die-down in 1861 implying about 15 cm of slip. Since the palaeoearthquake studies of Natawidjaja (2002) and Zachariasen et al. (1999, 2000) this region experienced the M_w9.3 earthquake of December 2004 which resulted in at least 2 m of slip on the subduction interface clearly representing the largest earthquake here for the last 250 years. Further slip just to the south of the 2004 earthquake site then occurred in March 2005 with an 8.7 magnitude earthquake. Indeed this latter earthquake resulted in substantial uplift/emergence of coral reefs and increased the land area of islands considerably.

Conclusion

The severe shaking of the lithosphere during earthquakes results in sig-nificant deformation of rock, sediment and soil layers. Fault terminations, sand blows and mud deformation features, landslides, anomalous sediment layers in lakes and destruction of ancient human built features provide us with an accurate long-term history of the frequency of these past events. Their magnitude can also be ascertained from the prehistoric evidence using the extent of surface displacement features, both along a fault and the vertical displacement of the ground, and the degree of offset of features such as aqueducts. The written and oral records of ancient civilisations also frequently record earthquakes because of the devastating impact they had on these societies.

The infrequent occurrence of earthquakes at a given location makes it difficult to obtain long-term records of sufficiently fine resolution to determine whether they have varied significantly in their frequency over time. The long-term records, however, can provide information on the average time interval between these extreme events and have often shown that the historical record is a poor reflection of the nature of the earthquake regime for a region. For example, parts of the Dead Sea Fault in Syria had been thought to be inactive and not likely to produce a severe earthquake in the near future. But the Meghraoui *et al.* (2003) analysis of the long-term record showed that an earthquake of greater than magnitude 7 is imminent and possibly overdue in this region. Once again, and like so many long-term records of extreme natural events, it is clear that we cannot rely upon short historical records to make reasonable predictions about the future impacts of earthquakes or, indeed, any natural hazard.

7

Landslides

Landslides occur when the shear stress imposed upon a slope consisting of sediment or rock (weathered and unweathered) exceeds the resistance of that material to movement downslope due to gravity. Shear stress is the force that promotes movement on slopes. Resistance to such movement is termed shear strength. On a stable slope the shear strength is greater than the shear stress. The stability of a slope, otherwise known as the margin of stability, is defined as the number of units of shear strength or resistance that exceed the shear stress. Slopes that are at the point of movement have little or no margin of stability. At this time, shear stress equals shear strength and the margin of stability equals zero. Marginally stable slopes are slopes that are approaching a margin of stability of zero (Crozier, 1986). The margin of stability can also be described as the factor of safety. Bell (1998) describes the factor of safety (F) as:

$$F = M_R/M_D \tag{7.1}$$

where M_R describes the resisting forces or the shear strength and M_D the disturbing forces or shear stress.

Crozier (1986) notes that slope instability is determined by both the internal state of stability (factors within the slope) as well as the magnitude of transient forces that normally occur outside the slope. Two common transient forces that lead to slope instability are climate and earthquakes. The climatic factors may range from dry to wet conditions, but high rainfall occurrences are generally associated with increased risk of slope movement in areas prone to such events. Heavy rainfall may affect the pore pressure in the material composing the slope, which in turn affects the internal factors that otherwise cause resistance to slope movement. Heavy and prolonged rainfall tends to cause more frequent larger, deeper landslides whereas shorter, shallower slides occur more frequently during short but intense rainfall episodes. Landslides resulting from earthquakes occur because of the sudden increase in shear stress (Bell, 1999).

Crozier (1986) suggests that the causes of slope instability can be divided into three groups.

(1) Preparatory factors that dispose the slope to movement. These factors make the slope susceptible to movement without initiating it. Slopes where preparatory factors are present can often be classified as marginally stable slopes.
(2) Triggering factors that initiate movement. The presence of these factors shifts the slope from marginally stable to actively unstable.
(3) Controlling factors that determine the nature of the movement in terms of the form, rate and duration of the movement.

The three main external factors that increase the chance of slope movement occurring are:

(1) steepening of the slope such as occurs during fault movements;
(2) removal of support from the toe or close to the toe of a slope (undercutting); and
(3) addition of material to the top of the slope or close to the top of a slope (overloading).

External causes of slope movement can result from human activities or from natural environmental processes, e.g. the removal of support at the toe of a slope. The removal of support can result from human activities such as excavation or environmental processes like stream or wave action. Any action that steepens the angle of a slope increases the likelihood of slope movement or slope failure. Internal causes of slope movement are usually considered to be long-term processes that include inherently weak materials, water content and pore pressure, decreasing cohesion between the materials and geological structures such as ancient slide surfaces, structures within the rocks such as fissures, joints and cracks, and rock layering dipping at an angle that is less than the slope angle. It is rare, however, that there is only one cause for a landslide.

Although a landslide can be triggered by specific events, long-term processes (internal factors) may cause a slope to become marginally stable (Abbott, 1999). For example, deep weathering of labile (easily weathered) rocks may cause a gradual build up of regolith or saprolitic materials on a slope. Under normal rainfall conditions the saprolite may remain stable. During an extreme or prolonged rainfall event the slope may fail along the contact between the saprolite and unweathered rock. After this failure occurs it will take some time (potentially

Table 7.1 *The role of vegetation in promoting slope (in)stability (after Bell, 1998)*

Mechanisms	Influence
Hydrological	
Foliage intercepts rainfall, causing absorptive and evaporative losses that reduce rainfall available for infiltration	Beneficial
Roots extract moisture from the soil, which is lost to the atmosphere via transpiration, leading to lower water pressure	Beneficial
Roots and stems increase the roughness of the ground surface and the permeability of the soil, leading to increased infiltration capacity	Adverse
Depletion of soil moisture may accentuate desiccation cracking in the soil, resulting in higher infiltration capacity.	Adverse
Mechanical	
Roots reinforce the soil, increasing the shear strength	Beneficial
Tree roots may anchor into firm strata, providing support to the up-slope soil mantle through buttressing and arching	Beneficial
Roots bind soil particles at the ground surface, reducing their susceptibility to erosion.	Beneficial
Weight of trees surcharges the slope, increasing normal and downhill force components	Adverse or beneficial
Vegetation exposed to the wind transmits dynamic forces into the slope	Adverse

several thousand years) before a similar size extreme rainfall event could cause a failure of the same magnitude. In other words, the long-term processes need considerable time to render the slope marginally stable by once again building up sufficient saprolite on the slope.

Changes in vegetation on a slope can increase or decrease that slope's susceptibility to a landslide. However, it is often the case that removal of vegetation usually leads to increased slope instability and higher erosion potential. Bell (1998) summarises the role of vegetation on slope stability as shown in Table 7.1.

The processes causing slope failure are complex. As such, several classifications of slope failure have been presented in the scientific literature. Many of these classifications are based on slope failure specific to one particular area and are not applicable to other sites where the geological and environmental variables differ. Due to the difficulties in defining the commonly used term landslide, several authors (Crozier, 1986; Bell, 1998) have advocated using the term slope movement for mass movements restricted to slopes. The term slope movement encompasses three forms of mass movement in rocks and soils, these being falls, slides and flows (Bell, 1998) (Fig. 7.1).

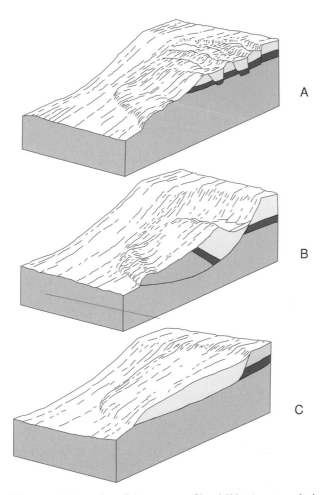

Figure 7.1. Examples of three types of landslide. A = translational block slide, B = rotational slump, C = earth flow (after Jibson and Keefer, 1994).

Falls occur frequently and generally describe an event where the moving mass travels mostly through air in free fall, saltation or rolling. In a rock fall, where a single block or blocks of rock detaches from the rock face, the individual components of the fall behave mainly independently of each other. Often there are no movements that precede such falls and the process itself is very rapid. Independent blocks increase in kinetic energy as the acceleration increases during the fall. The blocks accumulate at the bottom of the slope as a scree deposit. Freezing and thawing is one of the major processes that can result in rock falls. Toppling failure is also classified as a fall. The risk of toppling increases with increasing discontinuity angle. The discontinuity angle is the angle of a joint or fracture which separates the blocks from the rock mass. Toppling failures frequently occur on steep slopes of vertically jointed rocks.

Slides occur from shear failure along one or more surfaces that have reduced resistance against movement. The mass of soil and/or rock that moves during a slide may or may not experience deformation. The most common type of slide occurs in clay soils and results in an approximately spoon-shaped slip surface. These slides are called rotational slides and are generally characterised by backward rotation of the sliding mass (Fig. 7.1). They are also deep seated (the base of the slide occurs at some depth). These slides are generally associated with little deformation of the moving mass, and usually with the formation of only one or a few discrete blocks or masses. It is also common for successive rotational slides to occur at one location. This is a result of the scar left at the head of the slope being near vertical and unsupported until successive slides have stabilised the slope. Successive slides are defined as retrogressive slides and develop in a headward direction with a common basal shear surface where the individual planes of failure are combined (Bell, 1998).

Non-circular (non-rotational) slides occur in over-consolidated, weathered or unweathered clays. They occur along quasi-planar slip surfaces. These slides are called compound slides and are usually single events or progressive non-circular slides (Crozier, 1986). Transitional slides tend to occur in inclined, stratified deposits. Here, the slope movement occurs along a planar surface, usually a bedding plane. These slides occur when the mass is dislodged due to the exceedence of the gravitational force over the frictional resistance along the slip surface. Transitional slides are often slab slides where the slip surface is near parallel to the ground surface. These slides may occur on more gently inclined surfaces than rotational slides and also have the potential to be more extensive. Failure usually occurs in clays, quick clays and varved clays, and results from lateral spreading along bedding planes due to increased pore-water pressure. The increase in pore-water pressure generally occurs in a permeable zone at shallow depths that result in the mobilisation of the overlying layer. The mass generally moves over planar surfaces and may divide into a number of semi-independent slide units with complex movements that may include translational movements, rotation and liquefaction (Bell, 1998).

Slides include both rock and debris slides. Rockslides usually occur on steep slopes due to a gradual weakening of the bonds found within the rock mass. They are usually translational slides controlled by discontinuity patterns within the parent material. Again, freezing and thawing is one of the main factors controlling this type of slide. Rockslides may travel significant distances. Debris slides usually occur in the weathered parent material. Debris slides often occur in the form of mud slides which are generated by an increase in the water content along the contact between loose and/or saprolitic material and the firm underlying rock (Bell, 1998). Debris slides are usually composed of mud or other fine-grained sediments with coarser-grained material such as gravel and boulders carried in suspension; they move rapidly downslope.

The mass movements occurring in flows resemble that of a viscous fluid (Fig. 7.1). In this situation intergranular movement is more important than shear surface movement. As a consequence, slip surfaces are either not present or short-lived. The boundary between a debris flow and the underlying layer can be distinct or alternatively characterised by a zone of plastic flow. Although some water is necessary to induce most flow events, dry flows can occur, although these usually consist of rock fragments and result from slides or falls. Dry flows are usually rapid and short-lived and often occur in rugged mountain terrain. Wet flows usually occur in fine-grained sand and silt with or without coarser-grained material which is highly saturated. Hence they result essentially from an excess of water. Debris flows differ from mud flows in that the former have a higher percentage of coarse-grained material than the latter and usually occur after heavy rainfall or sudden thaw of frozen ground. Debris flows have a high density and have the capacity to carry large fragments such as boulders. Mud flows generally occur where large volumes of rapidly running water mix with debris to form a pasty mass. Mud flows have the capacity to move down shallow slopes. In these situations a forward thrust develops due to the undrained loading at the rear of the flow where the basal shear surface is inclined steeply downwards. Not only does this process result in further instability of the loading, it also results in increased pore-water pressure along the back part of the slip surface.

Earth flows result from increased pore-water pressure causing a decrease in shear resistance. Saturated flow material can result in the frontal part of the flow being recognised by a bulging lobe that may split into several tongues of flow that move in a steady rolling motion. Earth flows may form at the toe of rotational slides and represent a transitional movement between a slide and a flow. They may be rapid or slow moving depending on the water content of the moving mass. High water content usually results in fast earth flows, and the opposite is true of lower water contents. Earth flows, particularly ones with low water content, can continue to move for long periods of time (up to years) (Bell, 1998).

Magnitude of historical landslides

One of the largest landslides in historical times occurred in 1911 at Usoy in the Parmir Mountains of Russia. The landslide was generated by an earthquake estimated to measure about 7.0 on the Richter scale. The slide measured 2.5 billion m^3 or 2.5 km^3 and consisted mainly of soil and shattered rock. It buried the town of Usoy and its inhabitants, blocked the valley and dammed the Murgab River resulting in the formation of a large lake. The lake level soon rose, overflowed the landslide barrier and flooded Sarez, a village downstream (Bolt *et al.*, 1977).

The 1964 Alaskan earthquake also triggered large landslides. This earthquake measured 9.4 on the Richter scale. One of the resulting landslides, the Sherman Glacier slide, occurred about 140 km from the earthquake epicentre where one of the mountains, now called Shatter Peak, was shaken sufficiently well for 23 million m^3 of rock to break free from its crest. The falling blocks divided, gaining velocity as they fell and landed on a glacier forming a flat, lobe-shaped deposit only a few metres thick but which spread kilometres across the valley. It is possible that an aircushion may have formed and supported this landslide allowing it to travel the considerable distance across the valley. The aircushion could have formed when the slide material hit the glacier and a transverse oriented ridge at the bottom of the mountain that caused the slide debris to be launched into the air forming a relatively flat sheet of flying debris. As this sheet hit the snow and ice of the glacier, air could have been trapped between it and the glacier surface allowing the slide to move more easily than would have been otherwise expected. There were no casualties from the Sherman Glacier slide as the area was unpopulated (Bolt *et al.*, 1977).

About 200 000 people died in Kansu, China in 1920 when a large earthquake caused a massive landslide from the surrounding hills composed of loess (fine-grained wind-transported sediments often derived from glacial terrains). The earthquake broke the shearing resistance between the loess soil particles. The low permeability of this material resulted in air being trapped within the mass causing it to liquefy. Liquefaction is a common cause of landslides but it is usually water that becomes trapped within the material rather than air as occurred in the Kansu event (Bolt *et al.*, 1977).

The Portugese Bend landslide in California, USA was a relatively slow moving event. Urbanisation of the area during the 1940s and 1950s spread over former landslide deposits. Heavy rainfall in 1956 caused about 1 km^2 of this urbanised area to start sliding almost 20 m over a period of a few months on a slope of only 6.5°. The mass continued to move at a rate of about three metres a year and by 1980, 150 homes had been damaged or destroyed. The highway also had to be relocated resulting in an estimated total economic loss of $10 million. The LA County was held responsible for some of the damage as the area should have been left unestablished. The county had to pay about $1 million in damage (Bolt *et al.* 1977; Murck *et al.*, 1996).

Landslide impacts

Landslide events in the USA cost $1.5 billion annually in direct economic losses (Murck *et al.*, 1996). Twenty five to fifty lives are also lost each year to this geological process. While these losses in human lives appear great, they are considerably fewer in number than those occurring in less developed countries. This is, in the latter case, due to high-population densities, inappropriate zoning laws, a lack of scientific information regarding landslide events and the likelihood of such events occurring at any given site, and poor preparatory measures in case of such events. Although it has been pointed out that slope failure cannot always be predicted, planning schemes can be adapted to a specific site if economic prerequisites and scientific knowledge is available. Some of the largest losses of human lives due to landslide events are listed in Table 7.2.

Generally the cost of slope movements can be grouped into three categories. These are personal costs such as death and injury, economic loss by individuals or the general public and environmental damage. Environmental damage following a landslide event is usually short lived unless the event had a great range and extent, or the area in which it occurred is extremely unstable. The main impact on a slope following a landslide is loss of *in situ* biologic and pedologic material. Although these changes have traditionally been seen as a degradation of the habitat that prevent climax conditions from developing, it is likely that such slope movement events may play an important role in the maintenance of species diversity and rejuvenation. This can be compared to the effects of fire on many natural habitats. Larger slides have the potential to cause more severe and long-term impacts as the entire environment can be deformed and lose its biota. Slides can also derange natural stream patterns in an area as drainage pathways are altered, sometimes resulting in the development of poorly drained depressions and ponds. In more severe cases, as described in Chapter 6, landslides can cause the development of large lakes causing the submergence of previously terrestrial areas. Riparian vegetation and fish habitats may also be affected downstream from landslide events (Crozier, 1986).

PALAEOLANDSLIDES

Landslide events are often preserved for several thousands of years or more in a given landscape. This is because landslides involve the displacement of sometimes considerable volumes of rock, soil and sediment, and because these materials weather or break down relatively slowly. Records of past landslides, therefore, are often present in the stratigraphic record. They can also be recorded by other means, especially where they have impacted upon vegetation

Table 7.2 *Loss of human lives due to landslides*

Year	Place	Fatalities
1499	Kienholz, Switzerland	400
1515	Blenio Valley, Switzerland	600
1556	Shaanxi Province, China	830 000
1569	Hofgastein and Shwaz, Austria	287
1584	Yvorne, Rhone Valley, France	328
1618	Mont Conto, Switzerland	2430
1669	Salzburg, Austria	250
1741	Pennsylvania	22
1806	Goldau, Switzerland	457
1814	Boite Valley, Italy	300
1843	Mount Ida, Troy, New York	15
1881	Elm, Switzerland	115
1892	St. Gevais, France	177
1893	Trondheim, Norway	111
1903	Frank, Alberta, Canada	70
1920	Shaanxi Province, China	200 000
1936	Nordfjord, Norway	73
1938	Kobe, Japan	505
1945	Kure, Japan	1154
1958	Shizuoka, Japan	1094
1959	Hebgen Lake, Montana	28
1962	Nevados Huascaran, Peru	4000
1963	Vaiont, Italy	3000
1964	Anchorage, Alaska	114
1966	Aberfan, Wales	144
1966–67	Rio de Janeiro area, Brazil	2700
1969	Nelson Co, Virginia	150
1970	Nevados Huscaran, Peru	70 000
1971	St. Jean-Vianney, Quebec	31
1985	Mameyes, Puerto Rico	129
1998	Campania, Italy	180

or where they have altered the landscape, such as in the formation of lakes and deranged drainage patterns, and where they have provided new surfaces upon which vegetation, including lichens, can grow.

Lichenometry

Rock falls and other forms of landslide create deposits with fresh rock faces or surfaces. Lichenometry involves the determination of the age of lichens

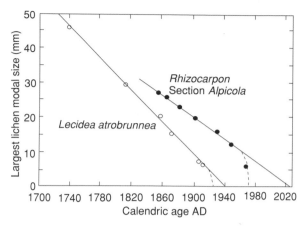

Figure 7.2. Growth rates of two lichen species growing on fresh rock faces in rock fall deposits, New Zealand (from Bull and Brandon, 1998).

growing on these freshly exposed rock surfaces. If the rate of growth of a particular species of lichen can be determined, then the measurement of the size of
lichens growing on a particular surface can provide an indication of the time
of exposure of that surface, and hence the age of the deposit and the landslide
event. Many studies have examined the rates of growth of various species of
lichens growing on tombstones whose age is usually inscribed as the year of
death of the individual. By examining tombstones of different ages, the rate of
growth, which need not always be constant for a given species, can be ascertained. With the same micro- to macro-climatic conditions the size of the same
species of lichen growing elsewhere can be measured and plotted against a curve
of known growth rate to determine an approximate age for a surface (Fig. 7.2).

Rock falls, rock avalanches and debris flows generate new rock exposures
upon which lichens can grow. Dating of the event or events that led to this
exposure is possible because lichen sizes record both the initial time of exposure of the rock surface upon which they grow and subsequent disturbances
that provide further fresh substrate where new lichens can colonise. The use of
lichenometry provides a direct date for the geomorphic event rather than an
indirect date like that provided by other methods such as radiocarbon dating.
In this sense, it is more precise and accurate than radiocarbon dating especially over the past few hundred years. Events only 2–4 years apart can be dated.
Bull and Brandon (1998) concluded that they could obtain event ages with 95%
uncertainty margins of ± 10 years in their lichenometry study of rock falls generated by earthquakes in New Zealand. But the range of possible ages using this
technique is largely determined by the frequency of reworking of the rock fall
deposit. As blocks are broken and covered by newly deposited blocks, or when

adjacent rock detritus and soils are moved, fresh substrate is colonised by new lichens. These periodic landslide events are usually recorded in the complex age differences of several colonies of lichen. The technique is limited where younger deposits cover older blocks, as the older surfaces cannot be sampled. Hence, bias against older deposits can occur and this can create difficulties when aiming to derive a magnitude–frequency analysis of events at a particular location.

Site selection, therefore, is important in the quest to obtain valid results, particularly where analysis of closely spaced events is desirable. Bull and Brandon (1998) note that several factors are important when choosing sampling sites. These include the diversity and frequency of geomorphic processes, lichen species and abundance, quality of thalli, substrate smoothness, the size of rock fall blocks and the ability to recognise old, stabilised block fields where lichen communities are not related to the times of substrate exposure because the first generation of lichens have died and been replaced. The latter generally provide minimal chronological information. First generation lichen communities are usually recognised by nearly circular, isolated lichens whereas lichen colonies that were established after the initial event are recognised by large thalli with highly irregular margins or a mosaic of thalli that have grown together. Slower growing lichens are usually preferable for dating prehistoric events.

There is no one specific method within the field of lichenometry. Traditional methods usually estimate the age of a geomorphic event by the size of the single largest lichen in a deposit or the mean size of the five largest lichens. The largest lichen, i.e. in terms of lichen radius or longest axis, is deemed to be the largest lichen found within a particular unit of time (e.g. 1 h). Bull and Brandon (1998) developed an alternative method based on populations of large lichens on several blocks within a deposit. The technique is known as the FALL or fixed-area largest-lichen method. The approach is based upon the style of sampling strategy employed and requires sufficient cobble and boulder size blocks in a sampling area to allow analysis of a number of individual lichen colonies. The FALL method measures the longest axis of the largest lichen in each of 100 or more sample sites of about the same size, typically 1 m^2. This sampling strategy allows the averaging of the effects of locally variable colonisation times and growth rates, taxonomic misinterpretation and inherited lichens that have merged to form composite thalli. Ideally, the sampling strategy should aim to minimise inherent variability in measurements by considering the sampling area and the density of lichen thalli in each area. Thus, sampling will ideally only include the largest isolated lichen in a number of sampling areas where the conditions for colonisation and growth are near identical but otherwise independent of each other. This is an ideal sampling strategy; however, actual variables may differ between sites depending on differing conditions. FALL

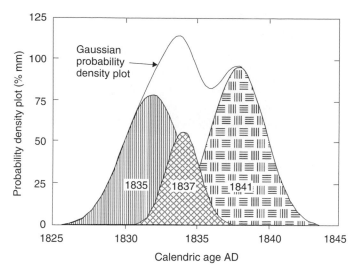

Figure 7.3. Probability density plot of times of regional rock fall events (from 47 lichenometry sites), New Zealand. Note that there are three distinct events here all separated by only 6 yr (from Bull and Brandon, 1998).

measurement is displayed most efficiently by using a probability density plot (Fig. 7.3).

Prominent peaks in the FALL distribution are viewed as a record of short-lived events that caused the reworking of parts of the deposit. Narrow peaks in the FALL distribution are typically associated with multi-event rock fall deposits with smaller standard deviations. Single event deposits have larger standard deviations and produce broader peaks. Multiple events can also be characterised by overlapping peaks. Probability densities are normalised to the same units, being percent per millimetre, to facilitate comparisons between density plots. Bull and Brandon identified three closely spaced events on New Zealand's South Island (AD 1835, 1837 and 1841). These landslides were related to earthquakes so the technique is also very useful for determining the long-term history of major earthquake events. The advantages of using lichenometry to date prehistoric earthquakes over more traditional methods, such as measuring fault movements, are that blind thrust faults and offshore subduction zones are very difficult to use for stratigraphic studies that show displaced rock and sediment units. Earthquakes associated with fault movements in these settings will, because of ground shaking, produce rock falls in mountainous areas.

It should be noted that most lichenometrical analyses are applied to an area without thorough knowledge of variations in lichen growth with local microclimate, altitude, temperature and precipitation, duration of snow cover and

Figure 7.4. Tree stem bend due to a landslide (from Lang *et al.*, 1999).

competition with other plants. There is also a lack of knowledge on variations related to substrate characteristics such as smoothness, lithology and degree of weathering. This lack of knowledge has limited the widespread and common use of lichenometry. It is important, therefore, that new calibrations of lichen growth are determined for each study site. Lichenometric dating is usually restricted to the past 500 years.

Tree-ring dating (dendrochronology)

Trees can record landslide events in two main ways – tree rings and stem tilting or distortions. Trees are subjected to stress during mass movements and this is recorded as a strong and sudden decrease in ring growth. Mass movements are also registered by tilting of the stem or an S-shaped stem (Fantucci and Sorriso-Valvo, 1999) (Fig. 7.4).

Tree rings are one of the most accurate methods for dating past landslide events (Fig. 7.5). The year of the event can be estimated and sometimes also the season. If cross dating of the samples cannot be conducted, age estimates for the

Figure 7.5. Tree ring scars from landslide impacts (after Lang *et al.*, 1999).

event can still be made. If the outermost ring in the sample is not preserved, the youngest remaining tree ring can give an estimate of the maximum age of the movement. A minimum age for the event can be determined by establishing the age of the oldest ring found in a tree established after the landslide event (Lang *et al.*, 1999). In this fashion the age of the event can at least be bracketed.

Tree-ring data are collected by taking cores through the tree trunk. The location of trees needs to be plotted on a topographic map, and the direction and degree of any stem tilting and the direction of the core sampling should be noted. Core samples are usually subjected to standard analysis procedures such as sample surface preparation, skeleton plots, cross dating and ring-width measurements with a micrometer to construct growth curves. The latter can also be cross checked statistically using computer software (Fantucci and Sorriso-Valvo, 1999). Tree analysis or dendrochronology is most suitable when applied to the analysis of the frequency and dynamics of shallow landslide events of a sudden and episodic nature, such as debris flows and rock falls, or rotational slides and mudslides that are reactivated within intervals of several years. Data recorded from continual or seasonal mass movements provide less accurate estimates of the time of the event (Lang *et al.*, 1999)

Cores can also be subjected to a method called visual growth analysis. This procedure is used to identify sudden increases or decreases of growth in the sample cores. A sudden decrease of ring growth is referred to as 'suppression', and a sudden increase of ring growth is known as a 'release' (Figs. 7.6 and 7.7). Growth anomalies found during this process are used to create a graph which is used to calculate an anomaly index (It) for each of the event-response cases (Fig. 7.8). The anomaly index considers the intensity of suppression as expressed by a coefficient of intensity suppression, the number of observed rings affected

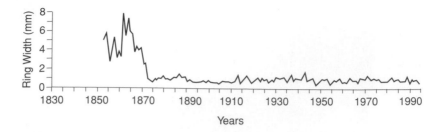

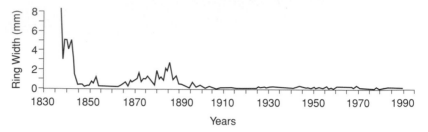

Figure 7.6. Growth curves for disturbed trees in northern Calibria, Italy. Strong suppression of tree-ring growth is evident after landslide events in AD 1844, 1871 and 1890. The suppression was so severe in these cases that the trees never recovered their normal growth. The two curves represent separate tree-ring sequences (from Fantucci and Sorriso-Valvo, 1999).

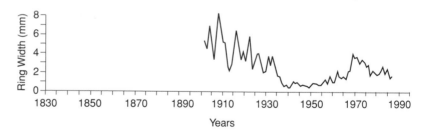

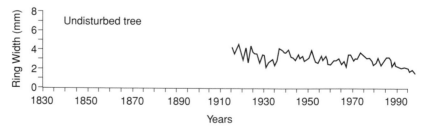

Figure 7.7. Growth curves as in Fig 7.6 except tree shown by first curve shows suppression of tree-ring growth following an event in 1931. Tree shows recovery from suppression after approximately AD 1965. Second curve shows undisturbed ring growth for tree not affected by landslides (from Fantucci and Sorriso-Valvo, 1999).

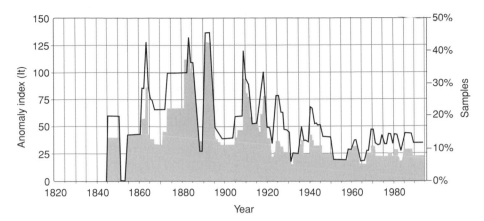

Figure 7.8. Visual growth anomaly index related to landslide area. The first registered disturbance occurred in AD 1845. The strongest events occurred between AD 1860 and 1896 as these affected up to 45% of the trees. The period between AD 1905 and 1950 shows a progressive decrease in trees affected from 40% to 23%, suggesting less intense events. The period between 1951 and 1995 shows least disturbance with affected trees not exceeding 16% (from Fantucci and Sorriso-Valvo, 1999).

each year and the total number of samples from each year. The following equation from Fantucci and Sorriso-Valvo (1999) describes this relationship

$$\text{anomaly index (It)} = \frac{\sum_{t=1}^{n} (Sup(x)t \times (Fx))}{\sum_{t=1}^{n} (N_{tot})^t} 100\% \qquad (7.2)$$

where, $Sup(x)t$ = the number of suppressions of each class (x) in year (t), Fx = intensity coefficient expressed by integers in a rank scale from 1 to 4 according to increasing intensity of suppression. These values correspond to slight, moderate, strong to very strong. (N_{tot}) = the total number of samples analysed in year t.

The visual growth analysis method places more emphasis on events that have resulted in strong suppression in trees (Fantucci and Sorriso-Valvo, 1999). Ages based on this technique sometimes need to be treated with caution as variations in ring width can also result from other factors such as climate, complex competition between trees as well as disease. The findings can only be deemed reliable if the samples exhibit strong replication and if other disturbing forces have been ruled out (Lang *et al.*, 1999).

Fantucci and Sorriso-Valvo (1999) used visual growth analysis to make comparisons between landslide events and records of seismic and hydrological (flood/rainfall) events. The results and comparisons were only moderately successful in connecting rainfall events and seismic events to landslide activity. Fantucci and Sorriso-Valvo (1999) suggested that the relationship between rainfall and landslides is considerably more complex than the one between seismically triggered mass movements. They found only two concurrences where rainfall and seismic activity occurred at the same time as a landslide (an earthquake between AD 1908 and 1910 and a 1910 rainfall event). While the identification of landslide events and their causative events was moderately successful, there was a clear distinction between the dendrochronological series from the unstable and undisturbed areas (Fig. 7.7). Hence, they were able to derive a record of active mass movements.

Ring-width patterns are also analysed for cross dating and to possibly say something about why the tree died or was subjected to impacts such as tilting. Ring-width patterns are usually studied from each tree in the data set. Cross dating is achieved by measuring all of the individual annual ring widths from multiple radii (two to four) for each tree sample and by making statistical analyses between samples. Essentially, cross dating involves the identification of a ring or set of rings with a common characteristic separate from the other rings. If characteristic X occurs in the outermost series of rings of a tree (Tree 1) and the same characteristic occurs in an inner ring of another tree (Tree 2) then the former tree (Tree 1) is older or started growing before the latter tree (Tree 2). The relative ages of trees in an area can then be determined by counting the rings either before or after the rings displaying characteristic X. If the absolute ages of any of the rings in a tree that contains a ring with characteristic X can be obtained, then accurate ages for any of the other rings can be determined. This of course is subject to the rings being annual growth features. The tree-ring patterns also allow the identification of the last growth ring produced before the tree died. The anatomy of this ring can provide clues to the cause of death of the tree (Jacoby *et al.*, 1992).

Jacoby *et al.* (1992) undertook a study of tree rings from trees submerged in Lake Washington, USA following three identified prehistoric landslide events. The presence of fire scars (the characteristic X in this case) in the sampled trees (Douglas firs) aided the cross dating of samples. These trees can survive repeated fires because of their unusually thick bark, but such fires can cause distinct trauma or scarring in subsequent rings. These scars provide excellent markers for the cross dating. The Jacoby *et al.* (1992) tree-ring analysis found that the trees had all died in the same year and season. It was suggested that the trees died between fall and early spring as the outer ring of all the trees was fully developed but without any initiation of the next year's growth. Radiocarbon dates for all three landslides overlap, as did the age of one of the trees. This evidence along with probable tsunami deposits and evidence for coastal subsidence all gave similar radiocarbon ages and led Jacoby *et al.* (1992) to conclude that the landslides in this area were triggered simultaneously by an earthquake. Although saturation from heavy rainfall and subsequent slope failure also had to be considered as a triggering mechanism, this was deemed highly unlikely as there are no historical records of such events occurring even during episodes of extreme precipitation.

Dendrochronology does not appear to be a reliable technique for determining the age of mass movement events where the failure surface of that movement is approximately parallel to the natural slope of the area. This is the case with planar slides where the mass movement occurs as a rigid block with no apparent deformation. In these situations the trees are only slightly tilted and the degree of tilting is not reliable evidence for the sliding process. Repeated landslide activity in an area can also produce complex dendrochronological estimates. In such cases, more thorough sampling and analysis methods are needed. These can include specially adapted sampling techniques that take into account the different stem bends, a systematic search for sudden changes in eccentricity or in reaction wood sequences and sampling of trees of different generations (Lang *et al.*, 1999).

Side-scan sonar

Side-scan sonar along with high-resolution seismic techniques have been responsible for the identification of giant submarine landslides. These landslides are instrumental in the evolution of many oceanic volcanic islands such as the Hawaiian chain, the Marquesas Islands, La Reunion and the Canary Islands.

Krastel *et al.* (2001) mapped 12 relatively young (2 million year old) giant landslides around the flanks of the Canary Islands (Fig. 7.9). Recognition of these landslides, their size and frequency are important for elucidating the risk from correspondingly giant tsunamis, which can be generated when these slides plunge into the ocean.

Interpretation of the pattern characteristics from side-scan radar images can give information about the composition of the submerged deposits. For example, speckled backscattering often depicts debris avalanches. Differences in the amount of backscatter can also provide further information about the debris avalanche as higher values of backscatter represent variations in the composition of the deposit. Side-scan sonar can also be used to determine the extent of the submerged landslide and the size of individual blocks. At El Hierro in the Canary Islands, side-scan sonar was used to study the morphology of the island's submarine flanks (Krastel *et al.*, 2001). The speckled backscatter pattern

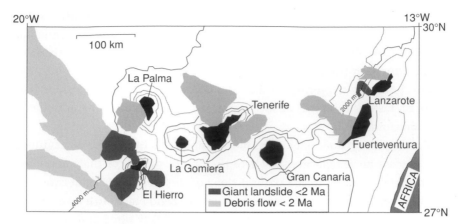

Figure 7.9. Giant landslides surrounding the Canary Islands (after Krastel *et al.*, 2001 and Carracedo *et al.*, 1999).

was assumed to represent hummocky materials formed from a debris avalanche. Differences in the amount of backscatter have also been used to identify the overlapping of debris avalanche deposits resulting from a variety of debris flows.

Side-scan sonar was also used successfully to determine and help date late Quaternary slope instability on the Faeroe margin east of the British Isles. The sonar images together with subbottom profilers and sediment core samples were able to identify recurrent mass movements on the Faeroe margin and the Faeroes slope of the Faeroe–Shetland channel. Sediment layers were separated through the roughness of the topography and the texture of the sediment deposit (Kuijpers *et al.*, 2001). Sediment echo sounding can also be used to create a profile of the seabed and detect individual blocks and allow for estimates of their size. As with side-scan sonar, the profile can also be used to estimate the extent of the landslide. Both these methods are frequently used together to provide a more complete picture of the nature of a deposit and its morphological characteristics (Krastel *et al.*, 2001).

By identifying the extent and size of submarine landslides these techniques help to ascertain the frequency and magnitude of possible associated hazards. Krastel *et al.* (2001) note that one giant landslide occurs on average every 125 000–170 000 years in the Canary Islands (Fig. 7.10), but this must be regarded as a minimum value. Four giant landslides have occurred since 200 000 years BP on the flanks of El Hierro and Tenerife resulting in one landslide every 50 000 years. More frequent smaller events have probably occurred as well. These landslides themselves represent a very serious hazard, especially if one were to occur today. Large lateral blasts due to the explosion of hydrothermal or magmatic systems may also be associated with the slides. The tsunamis generated

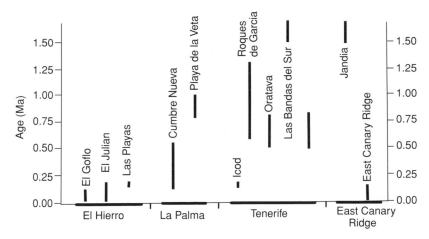

Figure 7.10. Ages of giant landslides around the Canary Islands (from Krastel *et al.*, 2001).

by the landslides would very likely cause serious inundation on neighbouring islands and may also travel thousands of kilometres across the Atlantic Ocean and possibly affect the Americas.

Stratigraphy

Repeated landslide movements will often result in an accumulation of landslide deposits towards the base of a slope. This pile is composed of individual landslide sediments which can be identified by their degree of weathering of component materials, the presence of soils, organic matter and other materials at the top of individual event layers, and also their relative position in the stratigraphy. Younger landslides normally overlie older ones so the sequence usually gets older with depth. Stratigraphic sequences can be identified in exposures or by drilling or boring cores from several sites within a study area (Elliott and Worsley, 1999).

Chihara *et al.* (1994) used stratigraphic techniques to determine the geohistorical development of the Tochiyama landslide complex in north-central Japan. They identified debris-avalanche and other landslide deposits. Three landslide layers are evident in the complex with the lower or older two composed of debris-avalanche deposits and the uppermost sequence composed of talus. Two distinct tephra layers separating these deposits allowed a chronology of these events to be developed. The oldest landslide layer was deposited between 46 500 and 25 000 years BP and the next upper layer deposited since AD 1361. These landslides were caused by local undercutting by streams, initially during

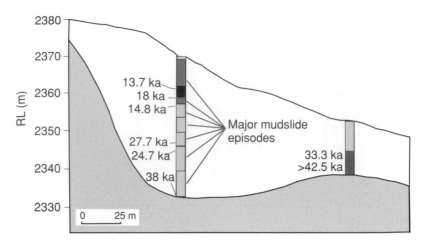

Figure 7.11. Mudslide layers in the Yakatabari Valley, Papua New Guinea (after Blong and Goldsmith, 1993). Note multiple event layers and radiocarbon dates showing initiation of slides around 40 ka. Dates from other boreholes show that the slide complex has remained active until at least 4 ka.

the last glacial maximum as sea levels fell and later by isostatic uplift of the surrounding terrain.

Blong and Goldsmith (1993) investigated the stratigraphy of the Yakatabari mudslide complex in Papua New Guinea. Boreholes in colluvial deposits at the study area allowed the construction of an isopach map of colluvial thickness and the identification of numerous phases of infilling of the valley due to landslides (Fig. 7.11). A number of landslide layers are evident in the 8–14 m deep boreholes drilled through the complex. Wood remains found in the sample cores and in near-surface exposures were radiocarbon dated. The wood would have been incorporated into the slides as trees were uprooted and destroyed during individual events. The complex, predominantly composed of mudslides, appears to have been active over the last 40 000 years. This is one of the few studies to ever identify such a long-lived landslide complex, but as Blong and Goldsmith (1993) note it is unlikely to be unique. It was not possible from this study to identify periods of more than a few thousand years when the mudslide was stable. These periods of stability are marked by the presence of buried soils. Buried soils represent periods of ground surface stability when the uppermost level of sediments undergoes weathering and is covered with vegetation. The soils will stop forming when another slide buries the existing ground surface. Studies such as this, where slides can be recognised and dated over considerable periods of time, highlight the persistent risks from landslides in the area. Blong and Goldsmith (1993) found no correlation between late Quaternary climatic phases and slide development and assume that this complex continues to become unstable over time due to episodic high-intensity rainfall events.

Aerial photography and field surveys

Examinations of aerial photographs combined with field studies are also useful ways to identify prehistoric landslides. These approaches have been used in a number of studies in the Himalayas (Hewitt 2001), the Canary Islands (Carracedo *et al.* 1999) and New Zealand (Crozier *et al.*, 1995). Carracedo *et al.* (1999) were able to identify giant landslides of the order of 200 km^3 by observing great landslide scars and deposits. Typical features identifiable from aerial photos include scarps, fans of landslide debris towards the base of slopes, dammed valleys forming lakes, asymmetrically thickened deposits and systems of pressure ridges or raised rims, the latter of which can sometimes be misinterpreted as lateral and terminal glacial moraines (Hewitt, 2001). Other features include independent streams or lobes of landslide deposits that have moved around obstacles, and up and down valley and into other valleys at tributary junctions. Hewitt (2001) also notes that sometimes a direct connection can be made or observed between boulder deposits on the valley floor and a detachment zone high on adjacent valley walls.

Crozier *et al.* (1995) used field surveys and aerial photographs to identify over 100 deep-seated prehistoric landslide events in the Taranaki region of New Zealand's North Island. They identified three separate groups of synchronously occurring landslides based upon the morphological characteristics of the landslide deposits and resulting landforms.

Table 7.3 lists the distinguishing features within two of these three groups. They include the degree of dissection of the deposit, soil development within the deposit and suppression of relief from volcanic air fall deposits. The older group of landslides show more advanced degrees of fluvial dissection and soil development (Fig. 7.12). The older group also have a more subdued relief because they have been exposed to, and mantled by, volcanic ash fall deposits for a longer period of time. The boundaries of the older slides were more eroded and it was rare to see them still damming lakes because streams had had sufficient time to breach the landslide barrier unlike the younger slides. Radiocarbon dates of organic fragments within the slides confirmed these observations showing that the younger slides occurred approximately 1200–1400 years BP whereas the older slides occurred around 30 000 years BP.

An interesting aspect of the Crozier *et al.* (1995) study was that they noted that no comparable landslides, i.e. multiple deep-seated landslides, had occurred during the 150 years of European settlement. They determined, using a variety of

Table 7.3 *Criteria for differentiating different age landslides in Taranaki region, New Zealand (after Crozier et al., 1995)*

	Group A (older)	Group C (younger)
Surface texture	Smooth	Rough
Boundary definition	Eroded; secondary slips	Distinct
Surface drainage integration	Well established	None
Fluvial dissection of mass	Up to 60 m	None
Valley dammed lakes	All drained	Many
Crown scarp lakes	Rare	Common
Dammed tributaries	None	Common
Soil	Volcanic ash mantle >15 cm A horizon	Residual soils with <15 cm A horizon
^{14}C dates (years BP)	31 500 ± 850	1380 ± 50
		1260 ± 50
		1250 ± 50
		1470 ± 250

Group B landslides have characteristics intermediate between groups A and C.

techniques and approaches (synchronicity of events, comparisons with modern analogues, spatial distribution and limiting equilibrium analysis), that the prehistoric landslides had all been caused by seismic events as opposed to episodes of heavy rainfall. Use of the limiting equilibrium analysis technique was critical to this assessment of the cause. The technique determines whether the slope geometry and the static shear strength conditions were susceptible to extreme hydrological conditions prior to the landslide event. If a factor of safety of 1.0 or less is found, the triggering is assumed to be hydrological. An analysis providing a factor of safety greater than 1.0 indicates that an added driving moment from seismic activity may have been necessary to trigger the movement. The technique is described by the following formula,

$$\frac{\text{sum of resisting forces}}{\text{sum of driving forces}} = \frac{C + (w^* \cos \beta - u) \tan \emptyset}{w^* \sin \beta} \tag{7.3}$$

where: C = the cohesion of the rock, w = the weight on the shear surface, β = the angle of the shear surface, u = the pore water pressure on the shear surface and $\emptyset$ = the angle of shearing resistance of the rock.

In all cases Crozier *et al.* (1995) found a factor of safety greater than 1.0 indicating that seismic activity triggered the landslides. They suggested that the triggering earthquakes must have had a minimum moment magnitude of 6.8–7.5.

Figure 7.12. Differences in morphological characteristics between young and older landslides in the Taranaki region of the North Island of New Zealand. (a) Hummocky terrain; (b) smoother and more deeply (fluvially) dissected terrain. Photograph courtesy of Professor M. Crozier, Victoria University, Wellington.

Statistical–modelling analysis

Modelling and statistical analyses can also be utilised to investigate the occurrence of past landslides. Jibson and Keefer (1994) used modelling techniques to determine the cause of landslides in the New Madrid seismic zone. Like the Crozier *et al.* (1995) study in New Zealand, Jibson and Keefer (1994) aimed to ascertain whether extreme hydrological conditions or seismic activity was the likely cause. The landslides are assumed to have occurred during the 1811–1812 earthquake based on historical notes and past literature. To test this, Jibson and Keefer (1994) analysed the slides under static or aseismic conditions and dynamic (seismic) conditions. The static conditions assume failure caused by changes in ground-water conditions in the absence of earthquake shaking. A broad range of physical or geotechnical factors are investigated in the field including the shear strength of the materials likely to fail, their plasticity, grain size distribution and unit weight. These factors are then incorporated into a numerical model. The dynamic analysis uses the Newmark displacement method which models a landslide as a rigid friction block of known critical acceleration, which is the acceleration required to overcome frictional resistance and initiate sliding on an inclined plane. Estimates of the minimum intensity of an earthquake event can also be made using this technique.

Jibson and Keefer (1994) concluded that groundwater conditions, even during extreme events, could not have triggered the investigated landslides. Based on prior knowledge of the earthquake in 1811–1812 and the described analysis it was determined that seismic shaking triggered the landslides. While this approach was undertaken in a known seismically active area, the method can also be applied in areas where the seismic conditions are unknown.

Conclusion

Landslides, like other geological extreme events, leave a distinct signature in the landscape. Block falls probably leave the most prominent imprint, simply because the blocks can be of substantial size. Of course they will eventually break down or weather and erode and/or be buried by other sediments but this may take considerable time. Until then lichenometry is a useful method, especially using Bull and Brandon's 'FALL' technique to establish the age of the event. Dammed lakes are more ephemeral but still a useful guide to past landslide events. Debris flows, mud flows and slumps are also obvious features but could be reasonably expected to last a shorter time in the landscape, at least as far as being an obvious event. Of course flows and slumps will remain in a stratigraphic sequence for longer but will require a trained eye for their detection.

Landslides and earthquakes are often intimately related in seismically active regions. Such is the case in New Zealand where the long-term history of landslides can also be used to infer the nature of the earthquake hazard. Likewise, landslides and tsunamis are related in some locations such as the Canary Islands. Here, side-scan sonar images of prehistoric massive slides now beneath the ocean surface show that these events would probably have created massive tsunamis. Dendrochronology has proved useful for reconstructing the frequency of many natural extreme events and landslides are no exception. Scars from landslide impacts and tilted stems help to identify the precise year that these events occurred. Apart, from tree rings, standard geological dating techniques probably do not provide the resolution required to determine detailed frequencies of landslide events. As such it is difficult to identify as yet whether landslides show any clustering over time. But like earthquakes, the prehistoric record certainly provides a good indication of the magnitude of the events possible in an area and, therefore, provides an invaluable aid to the assessor of risks from this natural hazard.

8

Volcanoes

Volcano and eruption characteristics

More than 90% of all volcanoes are associated with Earth's plate tectonic boundaries. Hot spots account for the remaining 10%. The most explosive volcanism occurs where plates converge, principally at subduction zones, but over 80% of magma released from the mantle by volcanic eruptions occurs at oceanic spreading centres (Abbott, 1999). Eruptions at spreading centres and over hot spots tend to be less violent than at convergent plate margins. This is principally because the former locations sit over the high-temperature asthenosphere where the rock has low densities of SiO_2. The pulling apart action by the oceanic plates here makes the upward movement of magma through fissures easier. The basaltic magma that rises at oceanic spreading centres releases gas easily and produces lava flows rather than explosive eruptions. This magma also has a high temperature with low viscosity. Because of these characteristics, these types of volcanic eruptions are the least hazardous to humans. Magma that is released at subduction zones incorporates subducted crust resulting in a higher water and SiO_2 content. It also has a higher viscosity (resistance to flow in a liquid) and explosive potential as gas is trapped in the magma and not easily released.

Hot spots are located where plumes of slowly rising mantle rock occur. This mantle rock is considerably hotter than the surrounding rock resulting in the mantle plume having a lower viscosity allowing it to rise. Tectonic plates ride over hot spots resulting in chains of volcanic islands marking the trace of the plate movement. The Hawaiian Islands are a classic example of such an occurrence and here the western islands of the chain are the oldest and least volcanically active as they have moved away from the hot spot. Hot spots can occur under continents or oceans, and they can also occur at the centre of tectonic plates or at spreading centres such as Iceland. Hot spots found over spreading centres have the potential to release more basaltic magma than hot spots under the centre of tectonic plates (Abbott, 1999).

Basaltic lava flows commonly exhibit three different types of texture: pahoehoe, aa and pillow lava. Pahoehoe lava is highly liquid and cools to leave a smooth, ropy surface. Aa lava flows are much more viscous and form a rough, blocky texture. Pillow lava occurs when lava flows reach water in lakes or the ocean causing it to cool rapidly into ovoid forms. Lava flows are often associated with basaltic magmas. Pyroclastic material, normally associated with the more acidic magmas, forms when gas trapped within the magma explodes resulting in the projection of debris into the air. This debris can contain magma and rock material broken into a wide range of sizes from dust, coarse-grained ash and cinders to blocks and bombs. Blocks consist of large fragments that are solid when airborne while bombs are composed of large fragments that are in liquid form when airborne. Coarser-grained material settles closer to the volcanic eruption whereas fine ash and dust can travel considerable distances.

There are three main characteristics that determine the form and structure of volcanoes. These characteristics are the type of magma feeder channel, the character of the material emitted and the number of eruptions that occur. The feeder channel may consist of either a central vent or a fissure. The material erupting from a volcano plays a significant role in determining the volcano's morphology. Stratovolcanoes form mainly from andesitic to rhyolitic magmas, being relatively high in silica, with high viscosity. These volcanoes are steep sided symmetrical forms composed of alternating layers of pyroclastic debris and lava flows. Shield volcanoes form from basaltic magmas which, because of their lower viscosity, flow for longer distances forming a volcano which is very wide compared to its height. Generally, volcanoes consisting mainly of pyroclastic material grow taller more rapidly than shield volcanoes (Bell, 1998).

Volcanoes can also be divided into monogenetic and polygenetic forms depending on whether they produce a single or multiple eruption, respectively. Monogenetic volcanoes have a central vent, are always small and have a simple structure where the eruptive centre moves on after an eruption. Monogenetic volcanoes usually occur in fields rather than individually. Polygenetic central vent volcanoes are more complicated and larger than monogenetic volcanoes. Stratovolcanoes are often polygenetic. The simplest form of these volcanoes is cone shaped, although the shape of such volcanoes may change due to several factors. These include migration and displacement of the vent to produce two or more summit craters and the formation of calderas. Calderas are formed when the volcanic peak collapses onto the magma chamber below. Polygenetic volcanoes can also consist mainly of lava flows such as those found in the Hawaiian Islands (Bell, 1998).

Bryant (2005) notes that explosive volcanoes produce five main types of eruptions. These are as follows.

(1) Strombolian eruptions which consist of moderate amounts of fluid lava material of all sizes that are thrown a few hundred metres up in the air. The lava is released in the air as bombs every few seconds. Events like this can continue for years. These volcanoes have symmetrical cinder cones and produce moderate lava flows. The magma is an intermediate stage between basaltic and acidic magma extrusions.

(2) Vulcanian eruptions expel large blocks of very viscous, hot magma that may be thrown more than 10 km in the air. In extreme cases viscous material can be thrown as high as 40 km. The activity is explosive and can span over periods of months. Vulcanian eruptions rarely produce flows. Tephra and block cones develop during the eruption. Between eruptions lava solidifies in the vents of the volcano and creates a plug. The strength of this plug determines the amount of pressure that can build up from underneath before another eruption occurs. Vulcanian eruptions are more acidic in nature than Strombolian eruptions.

(3) Surtseyan eruptions produce large dust clouds that can rise several kilometres in the air. These fine dust clouds are produced as hot, fluid magma comes into contact with seawater. Static electricity, resulting in lightning, occurs as fine-grained dust particles rub together in the rising cloud. The coarser-grained particles form rings around the vent rather than cones or ash sheets as is found in other types of eruptions. Due to the contact between magma and water, these eruptions are also violently explosive and are termed phreatomagmatic.

(4) Plinian (Vesuvian) eruptions cause more than 1 km^3 of magma to be projected up to 25 km into the atmosphere at speeds of between 600 and 700 m s^{-1} by a continuous jet stream and thermal expansion. The eruption becomes starved of material as the magma chamber empties and the magma changes in character. When this occurs the volcano usually collapses internally under its own weight to form a caldera. Such eruptions are even more explosive than Surtseyan eruptions.

(5) Peléean eruptions occur when the ash column from an explosive volcanic eruption collapses under gravity. Such an event can result in a debris avalanche of ash and hot gas that can travel down the slopes of the volcano at speeds up to 60 km h^{-1}. This debris avalanche is termed a pyroclastic flow or nuées ardentes. The debris left from such flows is called ignimbrite.

Non-explosive volcanoes produce two main types of eruptions.

(1) Flood lavas which are typically basaltic in composition. They are the least hazardous type of volcanic eruption.

(2) Hawaiian-type eruptions which are similar to flood lavas except that they can be faster flowing. Hawaiian eruptions also have the potential to produce tephra. These volcanoes can also produce large cones due to their aperiodic eruptions that build up successive deposits over time.

Impacts

There are six main hazards associated with volcanic eruptions. These are lava flows, basaltic and tephra clouds, pyroclastic flows and base surges, gases and acid rain, lahars (mud flows) and Jokülhlaups (glacier burst) (Bryant, 2005). Volcanic eruptions also have the potential to produce tsunamis. Volcanic eruptions can be associated with earthquakes and may be triggered by such events.

Lava flows

The velocity at which lavas travel depends upon their viscosity. The low-viscosity Hawaiian and Icelandic flows tend to move the fastest. They are difficult to stop, depending upon the type of lava as some, such as Pahoehoe, cool more quickly than aa lava. Attempts to cool lava down, build barriers or divert its movement have had mixed levels of success.

Ballistics and tephra clouds

Volcanic ballistics include up to boulder size accumulations of magma and hot rock material which can be projected relatively high into the air above the volcano. Tephra or ash can be projected tens of kilometres into the atmosphere and travel around the globe. This material can cover homes and farm land for many kilometres around the erupting volcano.

Pyroclastic flows and base surges

Pyroclastic flows occur as the ash column collapses under gravity during Peléean-type eruptions. Basal surges are produced during the lateral explosion of volcanoes and can pick up large amounts of dust in addition to ash from the volcano itself to potentially reach speeds of more than 150 km h^{-1}. Pyroclastic flows are avalanches of an exceedingly dense mass of hot, highly charged gas and fine particles. Pyroclastic avalanches can be very hot, having temperatures exceeding 1000 °C in extreme eruptions. As pyroclastic flows can also travel rapidly at speeds of up to 200 m s^{-1} a destructive flow is created that can overtop high topographic features.

Gases and acid rains

Gas can be emitted both during volcanic eruptions and from passive volcanoes and can be transported as acid aerosols, compounds absorbed on tephra particles or as salt particles. Large amounts of fluorine and chlorine are released into the atmosphere by passive volcanoes. In the presence of water during an eruption, hydrochloric, sulphuric, carbonic and hydrofluoric acids may develop. These gas particles, released as light acid rains, can stick to vegetation and kill crops or native vegetation. The acid rain can also cause skin irritations.

Lahars

Both primary and secondary lahars can occur during and after eruptions respectively. Lahars are mudflows that can be generated by pyroclastic flows or crater lake eruptions or saturation of fine deposited materials. These mudflows have the potential to destroy farmland through burying and pose a great threat to human lives depending on their size.

Glacier bursts

Glacier bursts are floods of heated water mixed with mud debris following the melting of snow and ice caps during a volcanic eruption. These muddy flows have the potential to threaten human lives. They may also flood the landscape and destroy crops and structures. Lahars can develop if these floods entrain sufficient mud.

Magnitude

The eruption of Tambora on the island of Sumbawa, Indonesia is regarded as the largest volcanic explosion of recent times (see Tables 8.1 and 8.2). In April 1815, it threw ash 20 km into the atmosphere and pieces of pumice up to 13 cm across landed over 40 km away. Tephra deposits were greater than 1.5 m thick at the base of the volcano and over 50 cm thick 150 km away. Tambora ejected over 100 km^3 of material and killed over 100 000 people. Fatalities occurred directly from volcanic activity and also because of starvation due to ash smothering agricultural land.

The eruption of Mount Vesuvius, in the Bay of Naples on the west coast of Italy in AD 79, was one of the largest volcanic eruptions ever witnessed (Cioni *et al.*, 2000). It was of the Plinian type. Plinian eruptions are labelled after the Vesuvius eruption as the eyewitness Pliny the Younger was the first to describe this type of eruption. After a day of large ash expulsion, a series of six surges and pyroclastic flows occurred in the area below the volcano. Almost 50% of

Table 8.1 *Largest volcanic eruptions of the last 10 000 years*

Year	Volcano	Country	km³ erupted
AD 1815	Tambora	Indonesia	100
AD 186	Taupo	New Zealand	100
3450 BP	Aniakchak	Alaska	50
6850 BP	Crater Lake	USA	30–40
6300 BP	Kikai-Akahoya	Japan	30–40
AD 1452	Kuwae	Vanuatu	32–39
3600 BP	Santorini	Greece	30–33
AD 50	Ambrym	Vanuatu	19–25
AD 1912	Katmai	Alaska	12
AD 1991	Pinatubo	Philippines	10
AD 1983–present	Kilauea	Hawaii	2
AD 1980	St. Helens	USA	1

Table 8.2 *Largest volcanic eruptions in prehistory*

Year	Volcano	Country	km³ erupted
28 Ma	Fish Canyon	USA	5000
1.3 Ma	Yellowstone	USA	1000
26 000 BP	Taupo	New Zealand	800
760 000 BP	Long Valley	USA	600

the cone collapsed to form a caldera with a diameter of 3 km. Ash falls were as high as 2–3 m and covered the towns of Stabiae and Pompeii. Although the town of Herculaneum was also affected by ash deposits and pyroclastic flows, its destruction was caused by a 20 m thick lahar generated by ground water expulsion as the volcano collapsed. Although most of the residents of Pompeii had evacuated, 10% were caught by the pyroclastic flow. Mount Vesuvius has had aperiodic eruptions up to the present. Major eruptions occurred in AD 203, 472, 512, 685, 787, five times between 968 and 1073, 1630 and 1906. Three hundred deaths occurred during the 1906 eruption due mainly to collapsing roofs when tephra up to seven metres thick fell on the northeast side of the volcano (Bryant, 2005).

The eruption of Krakatau in 1883 was also one of the largest explosive volcano eruptions in historic times. Its ferocity was largely due to the collapse of the caldera into the ocean. Krakatau lies in the Straits of Sunda between Sumatra and Java in Indonesia. Three months prior to a dozen Plinian-type eruptions, one vent had become active and thrown ash 1 km into the atmosphere. A few weeks later loud explosions occurred at 10 min intervals and a dense tephra cloud up to 28 km high rose over the island. The next day, another three explosions occurred. During the first explosion the 130 m high peak of Perboewetan collapsed to form a caldera that was immediately filled with seawater. The second explosion led to the collapse of the 500 m high peak of Danan. This resulted in more sea water being released into the molten magma chamber. The last explosion formed a 6 km diameter caldera where the peak of Rakata had been. This caldera was 300 m deep. As the third of these explosions occurred, an 80 km high, 18 km^3 volume, ash cloud formed. Ignimbrite deposits 60 m thick were deposited. Tephra fell over a total area of 300 000 km^2.

This eruption generated an atmospheric shock wave that travelled around the globe seven times, blocking the sun for two days downwind of the eruption. Solar radiation to the Earth was reduced by 13% as ash entered the stratosphere, and remained low for more than two years. Spectacular sunrises and sunsets could also be seen globally for up to two years as sunlight was reflected off dust particles. The Sunda Straits were blocked by large rafts of pumice. The collapse of the peaks, in addition to the pyroclastic surges and the shock waves, produced several large tsunamis, up to 40 m high, killing approximately 36 000 people. Tsunami waves were also experienced at various other sites globally such as the English Channel and Calcutta on the Ganges River. The sound of the last explosion was also the loudest sound ever heard in historic times and was heard 4800 km away (Bolt *et al.*, 1977).

The Mount St Helens eruption in Washington in 1980 was preceded by an earthquake triggering a large landslide. The movement of the landslide down the mountain caused a pressure drop in the gaseous magma. This, together with very hot ground water, produced an explosion and a pyroclastic flow that behaved as a low-viscosity fluid. The explosion resulted in the vent of the volcano opening, exposing the magma chamber and reducing the height of the peak by 500 m. This led to a 20 km high tephra cloud and another pyroclastic flow with temperatures up to 370 °C that attained a velocity of more than 100 km h^{-1}. Sixty people died during this eruption, mainly due to lahars (Bryant, 2005).

PALAEOVOLCANIC ERUPTIONS

Volcanic eruptions make an indelible impression upon people and communities. The truly large eruptions can become the focus of myths and legends. For this reason, long historical records of these hazards are preserved, especially the biggest ones, compared to other hazards. Archaeological evidence of the eruptions that impacted early civilisations also provides an excellent means by which to reconstruct the age and extent of prehistoric eruptions. Volcanic eruptions also leave distinct signatures in the landscape. Lavas, pyroclastic and ash deposits are preserved for thousands to millions of years. Vegetation communities are also affected by these events. Records of volcanic activity can thus be stored in pollen records. Each of these methods of reconstructing past volcanic activity is discussed below.

Archaeological evidence

Archaeological studies of prehistoric populations and settlements can be very useful in determining the nature and timing of volcanic events. Used together with other techniques such as stratigraphic investigations, petrographic analysis and geological dating it can help create a more complete picture of eruption events. Archaeological evidence that can aid the identification of prehistoric volcanic eruptions ranges from ceramic artefacts to the burial and preservation of entire cities by pyroclastic materials.

The burial of ancient buildings has provided evidence of the timing and intensity of a number of volcanic eruptions. The Cuicuilco pyramid in Mexico was covered by a 10–15 cm thick layer of ash and partially buried by a lava flow when the Xitle volcano erupted around 1670 years BP. In this instance, the archaeological evidence along with radiocarbon dating was able to show that the pyramid was still populated at the time of the eruption (Gonzales *et al.*, 2000). Also in Mexico, communities were forced to flee the area of Tetimpa in western Puebla, during a pumice fall event that buried buildings, activity areas and agricultural fields when the Popocatépetl volcano erupted between 200 and 50 BC. The event appeared to occur suddenly as many large and heavy household goods were left behind (Plunket and Uruneuela, 2000). A similar situation involving the burial of pre-Hispanic cultural features in the Canary Islands by volcanic debris helped to identify the age of the AD 1677 eruption of the San Antonio volcano. Here the Guanches peoples, thought to have originally come from North Africa, disappeared shortly after the occupation of La Palma Island by the Spanish in the last decade of the 15th Century. Pottery, petroglyphs, post-hole circles and other foundations of Guanche dwellings were buried by the eruption. While eruptions had occurred during or before occupation of the area by these peoples, the only volcanic units that overlie the Guanche remains were those of the 1677 eruption (Day *et al.*, 2000).

The Minoan city of Akrotiri was covered in volcanic material around 1500 BC following the eruption of Thera (Santorini). It should be noted that this is the date most commonly applied to this eruption although it is not certain as ages as old as 1648 BC have been suggested (Doumas, 1990). Thera lies just to the south of the Cyclades Islands, 110 km north of Crete in the Mediterranean Sea.

Archaeological and stratigraphic analysis following excavation of the city provided information about the nature of the disaster, the mechanism of the eruption and the sequence of events. The analysis also provided valuable information about the city itself, the standard of living and the architectural features. The stratigraphic record shows that the eruption was preceded by an earthquake prior to a small release of pellety pumice. After this event there was a pause in activity that was long enough for the volcanic crater to experience erosion. Five successive eruptions then covered the city under 30 m of ejecta. The eruption is also evident in the stratigraphic and archaeological record of other settlements in the eastern Mediterranean. It has even been suggested that the deposition of chaotic tephra in Egypt following the eruption can be connected to the Biblical Exodus episode where a great darkness blanketed Egypt for three days and was accompanied by a pillar of fire. However, this has not been established using comparative dating methods (Doumas, 1990).

The long-term effects of the Santorini (Thera) eruption on Minoan Crete have been recorded in the archaeological record. Changes are registered in architecture, storage and food production, artisan output, the distribution of prestige items, administrative patterns and ritual manifestations. The data suggest that the Santorini eruption resulted in significant economic dislocation in Crete. It is also assumed that the eruption resulted in a period of unrest and religious changes (Driessen and MacDonald, 2000).

Archaeological research has also found that prehistoric (and also more modern) societies used volcanic products where they were available. Volcanic ash is used in ceramics while obsidian is used both to make cutting implements and as semi-precious stones in jewellery. Other extrusive volcanic rocks are also used for building material, stone implements and abrasives. The petrography and age of such volcanically derived artefacts can provide information on the timing and nature of eruptions in a region.

In South America, volcanic ash found in ceramic artefacts from the Maya culture has been used to estimate the nature and age of prehistoric volcanic activity. Volcanic ash is abundant in ceramic utensils used throughout the Central Maya Lowlands (Guatemala) from the Late Classic Period (AD 600–900). Since volcanic ash is present in commonly used ceramic utensils such as cooking pots, as well as in limited and highly decorated artefacts, it can be assumed that it was easily accessible and abundant. The extensive use of volcanic ash, however, is difficult to explain as the nearest known volcanic ash field is more than 150 km away. Ford and Rose (1995) rejected the possibility that the ash could have been transported by draft animals over this distance and instead proposed that natural ash fallouts from eruptions in the Guatemalan highlands covered the lowland plains during the late Classic Period. They tested this hypothesis by undertaking petrographic analyses of the ceramics and a microprobe analysis of glass shards in the ceramics and compared these with the results of tephra and geochronological studies of the volcanic sequences in the highlands. The composition of major, minor and trace elements in the ceramics suggest that the ash used is consistent with tephra from the volcanic area in the Guatemalan highlands. The size analysis of the ash particles also indicated that the ash was windborne (volcanic loess material). These comparisons along with observations on the present occurrence of ash fall in the Maya lowlands following the AD 1982 El Chichon eruption suggests that ash may have been in abundance in the lowlands between AD 600 and 900. Comparisons of current wind patterns with historic ash falls in Guatemala are also consistent with the direction of airborne transportation and fallouts of the ancient deposits. The lack of such deposits in the area today, however, suggests that any such deposits have been subsequently removed most likely due to the rapid weathering processes in this tropical climate.

The volume of ash used in the Maya lowland ceramics was greater than 1400 m^3 yr^{-1} suggesting that there must have been considerably more outpouring of ash from volcanic eruptions at this time compared to that which has occurred over the past 550 years, or since historical records began in AD 1541 in Central America. This level of volcanic activity has not occurred during historical times and the archaeological record in this instance suggests that the Late Classical Period of Guatemala was dominated by active volcanism over a period of around 300 years or the time span over which these ceramics were made. Frequent volcanic activity also explains why settlements in the highlands were diminished or abandoned during the same period (AD 600–900). Ford and Rose (1995) have suggested that the Mayan lowland culture ceased at the end of this period as the volcanic activity in the highlands diminished leading to a reduction in soil fertility.

Besides ash used for ceramics, there are many volcanic products that have been used in antiquity that can be used to potentially derive a long-term history of volcanism for a region. Products such as volcanic rocks for use as stone tools, and for building and sculpture, for use as millstones and as additives for making cement can all be analysed to check for source area and geologically dated to determine a chronology of eruptions. The Romans used pumice to form hydraulic cements for use in marine and other architecture in the Hellenistic Period around the 4th Century BC. They used pumice in this way in the Pantheon in Rome to reduce its bulk density (Griffiths, 2000).

Although archaeological methods can be very successfully applied to investigate and date prehistoric volcanic eruptions, it can provide an even more accurate record of the impact of these eruptions on human settlements and culture.

Torrence *et al.* (2000) investigated the impact of volcanic eruptions on human settlements in the West New Britain region of Papua New Guinea. This region experienced up to 13 eruptions from the Witorio volcano and four eruptions from the Dakataua volcano during the mid- to late-Holocene (5000–1000 years BP). These volcanic episodes resulted in the regular abandonment of settlements, the intensification of subsistence agriculture and a punctuated trend in the use and development of stone tools.

Volcanoes and mythology

It is entirely possible that many myths from various cultures are based upon real extreme events. This seems reasonable if these events had a profound impact upon those peoples. Volcanic eruptions are clearly dramatic events and of course have been devastating to many cultures. Santorini and Vesuvius are classic examples. It is interesting then that myths from the ancient peoples of the Mediterranean region appear as if they could be based upon major volcanic eruptions of the past. Hesiod's *Theogony*, for example, is a record of the lineage of the Greek Gods and their struggles for power. It is the earliest surviving document of Greek literature devoted mainly to mythical topics. Greene (1992) has suggested that major volcanic eruptions may have been the inspiration for the battles between the gods, and particularly the sequence of events in these battles and their appearance, sounds and effects on the physical world. The battle between Zeus and the Titans, possibly describes the Santorini eruption at Thera around 1500 BC. The battle between Zeus and the monster Typhoeus may also be an accurate description of the Mount Etna eruption of 735 BC. If these myths were based upon real extreme events then they may have survived as oral stories or descriptions for approximately 700 years until they were first written. However, if they did survive for this length of time, the accuracy of the description of the actual event may be questionable. It is also uncertain whether the particular eruption described can be distinguished from other volcanic eruptions on the basis of qualitative, colloquial and figurative descriptions in the mythology and the geological evidence (Greene, 1992).

Geological records of prehistoric volcanic eruptions can at least help to better ascertain the nature of the relationship between the myths and actual events. The battle between Zeus and the Titans does, for example, follow a logical sequence of a volcanic eruption. Table 8.3 compares the sequence of events in Zeus's battle and that likely to have occurred during the eruption at Thera. Not only does the physical and chronological sequence of the eruption follow the physical and chronological sequence of the battle, but the time around 1500 BC when the volcano actually erupted fits with the assumed time of the battle.

Table 8.3 *Comparison between the battle of Zeus and the Titans and the likely sequence of events during the eruption at Thera (from Greene, 1992)*

Hesiod's description (Zeus vs Titans)	Eruption of Thera
A long war	Premonitory seismicity
Both sides gather strength	Increasing activity
Terrible echoes over the sea	First phase explosions
Ground rumbles loudly	Tectonic earthquakes
Sky shakes and groans	Air shock waves
Mt Olympus trembles	Great earthquakes
Steady vibrations of the ground	Earthquakes
Weapons whistle through air	Pyroclastic ejecta
Loud battle cries	Explosive reports
Zeus arrives: lightning, thunder, fields, forests burn	Volcanic lightning, heat of ignimbrites
Earth and sea boil	Magma chamber breach
Immense flame and heat	Phreatomagmatic explosion
Sound of earth/sky collapse	Sound of above
Dust, lightning, thunder, wind	Final ash eruptions
Titans buried under missiles	Collapsed debris

Stratigraphy and tephrochronology

Stratigraphic methods are commonly used to identify, analyse and date episodes of volcanic activity. Stratigraphic units can include lavas, pyroclastic deposits, tuffs (sedimentary deposits eroded from volcanic rocks), and pumice and ash fall deposits. The stratigraphy can be analysed where streams have cut into the sequence and also from cores. Grain-size, lithological and geochemical analyses of individual units help to determine the origin and age of eruption events. The stratigraphy of ice cores in the Arctic region and Antarctica can even reveal the age and sometimes the location of large distant eruptions when ash has entered the upper atmosphere. Stratigraphic analysis is probably the most reliable of all methods in accurately reconstructing the long-term history of volcanic activity of an area. A variety of dating techniques can be used including radiocarbon, isotopes and luminescence depending upon the quality of the preserved materials (Lecointre *et al.*, 2002).

Tephrochronology is the study of tephra layers in stratigraphic sequences. Comparison of the tephra's chemical components with modern analogues helps to identify the specific volcano or the volcanic field from which the layer originated. Dating of the layer, or those immediately above and below, depending upon the nature of materials in the stratigraphic sequence, can determine the age of the eruption. Tephra layers can mark both local and distant volcanic eruptions. In the latter case, ash is often projected into the upper troposphere and as a consequence can circle the globe. Comparisons between identified tephra layers and possible source volcanoes are made by comparing the concentrations of major oxides within the layer. Tephra shards are often used for this type of analysis. However, locating the source volcano by the geochemical signature of these shards is not necessarily straightforward and extensive dating of the stratigraphic sequence is often first required in order to gain an idea of which volcanoes around the globe may have erupted at that time. Even then, problems can occur as dating techniques have an inherent level of inaccuracy. For example, a tephra layer marking a major volcanic eruption in AD 1459 occurs in ice cores in Antarctica and Greenland. It was probably one of the largest volcanic eruptions globally over the past 700 years. The eruption most likely occurred in Vanuatu when the volcano Kuwae destroyed an island (Australian Antarctic Division, 2004). Historical records and tree rings suggest that this eruption occurred in AD 1453, six years earlier than that suggested by the Antarctic ice cores. But even when this is taken into account, along with potential errors in dating the ice core, the earliest date for this eruption (based upon the ice core record) is AD 1456. Such potential differences in determining the age of a specific eruption event will become even more exaggerated when using dating techniques such as radiocarbon or luminescence which have considerably larger uncertainty margins than tree-ring and ice-core dating. And this is especially so where possible eruptions are closely spaced in time. Another potential problem encountered during the correlation of tephra layers with possible eruption sources lies in the analytical methods applied to the micro-analysis of tephra shards. It is often difficult to compare data generated by different operators on different analytical machines which can result in reduced robustness of the data analysis.

Charman and Grattan (1999) suggest that it is possible to overcome these difficulties by using discriminant function analysis (DFA). This type of analysis uses a classification model based on a reference data set containing the major oxides from known tephras. The classification model consists of a series of discriminate functions that can be plotted graphically on discriminate function axes and applied to the unknown tephra. The unknown tephra is then classified into one of the known groups of tephra and is given a known probability for misclassification.

Tephras can also be identified by their magnetic characteristics, a technique known as magnetostratigraphy. The type and concentration of magnetic minerals in tephra layers depends on the chemical composition of the tephra. Different types of glass-encased magnetite grains and ferrimagnetic components can be identified by their magnetic components and magnetic susceptibility. These magnetic characteristics reflect the chemical composition of the source material and the temperature regime that existed during the eruption from which individual tephras were ejected and the subsequent chilling of the magma. One great benefit of using magnetic susceptibility to identify tephra horizons is that these layers can be identified in cases where field identification is difficult or even impossible. The magnetic properties of the tephra can give information about the type, concentration and grain size of the magnetic mineral and thus temperatures and cooling processes following the eruption (Gonzales *et al.*, 1999).

Tephra magnetostratigraphy is quick, easy and economical compared to geochemical methods. Along with ice cores, tephra layers can also be identified in soils and sediments. Soils will also show magnetic characteristics, but individual tephra layers can be discriminated because the back-field isothermal remnant magnetizations (IRMs) highlight the presence of different types of magnetic minerals for the soils (magnetite) and tephras (paramagnetic minerals). Curie temperature curves also differ between magnetite and paramagnetic minerals and these curves can be used to further help differentiate between soil and tephra layers (Gonzales *et al.*, 1999).

Beget *et al.* (1994) used magnetostratigraphy along with geophysical and geochemical techniques to identify tephra layers in sediments in Skilake Lake, Alaska. Sediment cores from the lake floor contained at least nine separate, thin layers with anomalously high magnetic susceptibility. Petrographic analysis showed these horizons consisted of volcanic ash. These layers were not easily visible to the naked eye as the sediment cores consisted of several dark and light layers that were usually between 0.1 and 0.5 cm thick. Volcanic glass shards were separated from the tephra layers and individually characterised using an electron micro-probe (EMP). Each grain was analysed individually, which avoided the problem of misidentifying detrital contaminates and density fractionation during transport and emplacement. The source volcanoes for these tephra layers were identified by comparing the separate tephra layers with reference geochemical data sets from previously known and dated tephras. Four volcanoes were identified as the source of the nine tephra layers. Historic eruptions that deposited tephra layers in the lake sediments occurred in AD 1912, 1902 and 1883, and prehistoric eruptions occurred around 250–350, 300–400, 350–450 and 500 years ago (Fig. 8.1). Even older tephras were identified lower in the core but these were difficult to identify and date. The tephra layers and historical records show that volcanoes in this region erupted every 10–35 years during the 20th Century and ash falls accumulated in the lake at least once every 50–100 years over the past 500 years. The ash fall deposits in the lake alone probably underestimate the volcanic history of the area. Eruptions may not have been recorded if they did not have a significant ash deposit or if the ash was transported in an opposite direction. Comparison with records and chronologies for

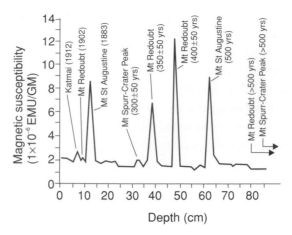

Figure 8.1. Tephra chronologies from Skilake Lake, Alaska over the last 500 years (from Beget *et al.*, 1994).

other areas, and using other methods, can provide a more complete picture of prehistoric volcanic eruptions (Beget *et al.*, 1994).

Stratigraphic sequences can also reveal the nature of volcanic eruptions and in particular, the relative violence of those eruptions. Siebert *et al.* (1995) examined tephra, debris avalanche and pyroclastic layers to reveal a history of summit collapse at Augustine volcano, Cook Inlet, Alaska. Here, 100–500 m of the volcano summit collapsed three times over the past 500 years during major volcanic eruptions. The presence of debris avalanche and lateral blast deposits at Mt Augustine also suggests that tsunamis may have been a hazard in the past and will again in the future. The avalanche and lateral blast deposits on Augustine Island, and also the nearby West Island, indicate high-frequency, cyclic collapse and regrowth of the volcano. The deposits indicate a recurrence interval of about 150 years and at least 11 collapse episodes have been identified for the past 1800–2000 years (Siebert *et al.*, 1995). The regularity and severity of these events over this time suggests that similar events will almost certainly occur in the near future and could cause considerable impact upon local human populations.

Isotope and radiocarbon dating

Direct dating of lavas and pyroclastic deposits is usually undertaken using Argon–Argon (^{40}Ar–^{39}Ar) and Potassium–Argon (K–Ar) dating. However, these techniques are not useful for estimating chronologies for young volcanic formations (<1 Ma) as these isotopes have relatively long half-lives. This is particularly the case for arc volcanoes as these usually have low potassium (K) content. However, dating can be accurate where the sample selection and preparation is carefully done and where the analytical procedures are optimal. Lava formations from the Taupo Volcanic Zone in New Zealand have been isotopically dated (Gamble *et al.*, 2003). Dating of select crystalline groundmass separates provided the most accurate dates. Less precise dates came from whole rock samples, and pure plagioclase separates. Gamble *et al.* (2003) found that most of the K in the crystalline samples was held in glass or groundmass μ-scale sanidine crystals formed during late-stage groundmass crystallisation. Groundmass samples could then be classified into four categories: glassy, moderately glassy, moderately crystalline and crystalline. The results and precision of the dating varied greatly between samples and the success of dating events younger than 20 000 years BP was limited (Gamble *et al.*, 2003).

Radiocarbon dating cannot date volcanic materials directly as it relies upon the decay of ^{14}C in relation to the amount of ^{12}C in a sample of a formerly living organism. However, radiocarbon can be used to date organisms that have met their death due to the volcanic eruption and hence mark the actual time of the volcanic event. Charcoal is useful in this context, for it can form when vegetation is buried under advancing lava flows. Lava flows are of course very hot and they will burn vegetation to charcoal relatively quickly as the lava front overrides that vegetation. Charcoal will burn to ash when temperatures exceed 500 °C. This is the case whether oxygen is present or not. However, where oxygen is present, the charcoal will burn rapidly once this temperature is exceeded. Although much of the charcoal is burned to ash immediately, some charcoal can be preserved when the availability of oxygen is restricted and the temperatures are below 500 °C. While lava flows will restrict oxygen flow from the atmosphere to charcoaled organic material, oxygen derived from highly porous, thick soils covered by lavas can restrict the preservation of charcoal. Entire forest areas in Hawaii have been covered by lava flows with no evidence of underlying charcoal because of oxygen rich soils. However, in the absence of these soils charcoal can be preserved and used for radiocarbon dating of a relatively young or recent lava flow (Lockwood and Lipman, 1980).

While some difficulties are inherent in the recovery of charcoal from beneath lava flows, charcoal has a higher chance of survival when lahars bury vegetation. Forests in the path of these flows can be destroyed and buried and not be subject to the searing temperatures associated with lava flows. Cameron and Pringle (1991) dated forests buried by lahars generated during the eruptions of Mount Hood in Oregon to determine the recent eruption history of this volcano. Lahars and pyroclastic flows travelled many kilometres down rivers leading from this volcano to bury coniferous forests growing on the valley floors and lower slopes. Some trees were snapped off by high-velocity pyroclastic flows which may have attained speeds of 140 km h^{-1} and the trunks were transported and buried further down slope. Where slopes were shallower and flow velocities lower, the flows buried the roots and lower trunks of the trees and the remains of these forests remain *in situ*. Six buried forests have been found in this region resulting from three separate eruption events over the past 2000 years. Trees buried by these separate eruptions returned radiocarbon ages of 1700 ± 700, 560 ± 150 and 455 ± 135, 260 ± 150 and 185 ± 120 radiocarbon years BP. Two of these eruptions clearly occurred over the last 500 years with the last occurring from approximately AD 1760 to 1810.

Desktop studies

Desktop, as opposed to field-based, studies can also be useful in reconstructing the volcanic history of a region (Crattan *et al.*, 1999). Maars are volcanic craters formed from phreatic or phreatomagmatic explosions. Carn (2000) suggests that the morphology of prehistoric maars can be used to determine the relative age or 'freshness' of the formation based upon their diameter–depth ratio. Young maars have a diameter–depth ratio of 4–6. This value increases with age as erosion processes widen and fill the crater. Older maars have a diameter–depth ratio of about 10. Carn (2000) used this technique to estimate a relatively young age (14 000–40 000 years BP) for the maars and other volcanic activity in the Lamogan Volcanic Field (LVF), Indonesia. The crater diameter can also be used to estimate the ejecta volume from the eruption. Carn (2000) used the total ejecta volume (V_E) for each maar (seven in all) to constrain the volume of juvenile magma (V_J) by subtracting the crater volume (V_M) from the dense rock equivalent (DRE) of (V_E). Conversions to and from DRE were achieved using an ejecta density of 1.3 g cm^{-3} and a juvenile tephra density of 1.8 g cm^{-3}. The approach assumes that the crater has not been significantly widened since formation. This is reasonable when considering the relatively small diameter–depth ratios of maars. (V_M) ratios are subject to errors where the maars contain lakes; however, this is not an issue where the lake depth is known (Fig. 8.2).

Ejecta calculations can also be applied to cinder and spatter cones with well preserved craters. Volumes of exposed lava flows (V_L) are calculated and used in estimations of total cone eruption volume (V_{tot}). The cone basal diameter to height ratio (W_{co}/H_{co}) and slope angle (θ) can then be used to estimate cone age (Fig. 8.3). Existing data suggest that W_{co}/H_{co} increases and θ decreases as cones are eroded over time. Relatively fresh cones have indices of about 5–6. Low estimates of W_{co}/H_{co} can in some cases be a result of underestimates of W_{co}, which is poorly constrained for older cones that may have been buried by younger deposits. It should be noted that without radiometric dates, the age estimates are relative to younger and older formations in the same area. Also

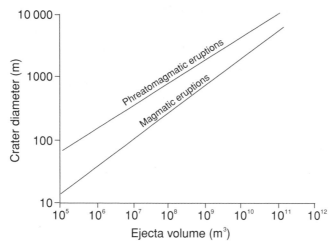

Figure 8.2. Maar crater diameter versus ejecta volume (from Sato and Taniguchi, 1997).

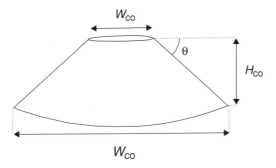

Figure 8.3. Typical cone morphology showing parameters used to calculate the cone age (after the Carn, 2000).

the weathering rates of different climates may affect these calculations. For example, cones in arid environments can be morphologically fresh even when 15 000–100 000 years old, whereas cones in the humid tropics start to erode only a few decades after formation. Cones in the Lamogan Volcanic Field are as young as a few centuries in age.

Time-averaged eruption rates (Q) can also be calculated. The interpretation of Q is dependent on the duration of the eruptive area in question. Calculations over longer intervals are less susceptive to weighting by periods of intense activity and eruptive rates averaged over longer periods will be a better indication of long-term volcano behaviour. Such calculations, however, require extensive eruption records. The ratio of intrusive or extrusive magma is also an important factor when calculating Q. Volcanoes that grow mainly by magma intrusion will have low Q unless the supply rate is particularly high. Estimates of eruptive volumes and Q can also be used to give rough estimates of age. In the LVF, the total volume of erupted magma is estimated to be approximately 12 km^3. This estimate must be regarded as a minimum value as eroded material is neglected in the calculation. Some cinder cones and maars, and most of the tephra produced during explosive eruptions, are also usually neglected. The age of the LVF, assuming a constant Q over the lifetime of the volcanic field and no major hiatus in volcanism, agrees closely with estimates based upon maar morphology (14 000–40 000 years BP).

Pollen records

Pollen records, in association with geologic records, can help to identify major episodes of volcanism. Stratigraphic analysis incorporating tephrostratigraphic records in the North Island of New Zealand show that the region has experienced a violent volcanic history throughout the Quaternary. The main volcanic centres are Okataina, Taupo, Maroa and Mayor Island. The Kaharoa tephra layer dated to 665 ± 15 years BP highlights one of the more recent of these major eruptive episodes. The tephra layer was ejected from the Tarawera volcano in the Okataina Volcanic centre and consists of a sequence of fine ash overlying coarse ash. Giles et al. (1999) found that this eruption and others like it have had a profound effect on New Zealand vegetation communities. Immediately following deposition of the tephra layer, the pollen record indicates significant disturbances. The levels of broken and degraded pollen grains increase and there is a sharp rise in pollen from species that are assumed to be invaders following the disturbance. There is also an increase in pollen levels from some tree species that thrive on fresh volcanic surfaces. Likewise, there is a decline of previously dominant species following the deposition of the tephra layer. Other mechanisms for the dramatic change in vegetation communities based upon the pollen record cannot specifically be ruled out; however, the scale and nature of the change strongly suggests that volcanic eruptions are the most likely cause.

Conclusion

Volcanic eruptions leave an indelible imprint in the landscape, both human and natural. The classic studies of Pompeii and Herculaneum after the eruption of Vesuvius are well documented, but other investigations such as that by Ford and Rose (1995) where they examined volcanic material within pottery made by the Mayans in central America also throw light on the nature of volcanism in that region. Likewise, mythology offers glimpses into major volcanic eruptions of the past. Lavas, pyroclastic flows and tephras can be preserved for millions of years. Stratigraphy, geological dating, isotope analysis and magnetostratigraphy have become some of the main tools for unravelling the long-term history of volcanic eruptions. The evidence can also occur much further a field than close to the erupting source. Fine-grained ash layers are evident in ice layers in Antarctica and the Arctic and these markers represent truly large eruptions where ash has been projected into the stratosphere and encircled the globe. Pollen studies of the demise of vegetation communities due to volcanic eruptions also offer an interesting form of proxy evidence as do the desktop studies of the size of volcanic maars as a guide to the magnitude of prehistoric eruptions.

9

Asteroids

The Titius–Bode rule states that the radii of the planetary orbits in our solar system increase geometrically with increasing distance from the Sun. This rule suggests that there is a planetary gap between Mars and Jupiter (Encrenaz *et al.*, 1990), or in other words that a planet should have once existed here. On the 1st January 1801, the missing planet was discovered by Italian astronomer Giuseppe Piazzi. He named it *Ceres* (Jones, 1999). It was soon discovered though that *Ceres* was not a planet as it is too small; *Ceres* is a minor planet or what is now known as an asteroid. Over the following years a number of other asteroids were discovered in the gap between Mars and Jupiter. This gap is now called 'the main asteroid belt' (Encrenaz *et al.*, 1990).

The word asteroid means 'resembling a star' (Jones, 1999). They are metallic, rocky bodies without atmospheres that orbit the Sun. They are too small to be classified as planets, hence the name 'minor planets'. They are largely confined to the main 'doughnut-shaped ring' asteroid belt between Mars and Jupiter from approximately 2 to 4 AU (AU = Astronomical Unit which is the distance from the Sun to the Earth and in this case 2–4 AU = 300–600 million km) (Jones, 1999; NASA, 2004). Tens of thousands of asteroids congregate in the main asteroid belt, but not all are confined to this belt for some pass much closer to the Earth; these are known as the near-Earth asteroids (Jones, 1999). Asteroids have a diameter of greater than 1 km and indeed can be larger than 100 km across. The largest asteroids to be discovered are *Ceres, Pallas, Juno* and *Vesta*, respectively (Encrenaz *et al.*, 1990). *Ceres* has a radius of 457 km, amounting to about one-third of the total mass of the entire main asteroid belt (Taylor, 2001). About 8000 asteroids have been observed to date and their orbits have been determined and categorised. Of these, 238 are greater then 100 km across (Jones, 1999). It is likely that most of the asteroids greater than 100 km have been discovered, but only a tiny proportion of the smaller asteroids have so far been observed. It is estimated that there are in the order of 10^9 asteroids greater than 1 km across (Jones, 1999; Taylor, 2001). Asteroids are classified by their surface composition and, in particular, their albedo (Encrenaz *et al.*, 1990; NASA, 2004). An asteroid is different from a comet, as a comet develops a thin atmosphere when it is within 10 AU of the Sun whereas asteroids do not (Jones, 1999).

Cosmic origins of asteroids

Asteroids are thought to have been derived from the planetesimals and embryonic systems that were in the space between Mars and Jupiter during the early formation of the Solar System. Asteroids are small condensations from the primitive solar nebula that were unable to form into a single body due to the gravitational instability caused by the presence of Jupiter (Encrenaz *et al.*, 1990). Interactions between asteroids and Jupiter, and the asteroids themselves, have resulted in collisional fragmentation of the asteroid population over the last 4600 million years (Jones, 1999). These collisions have also resulted in a substantial loss of material from the asteroid belt over time; there was about 10^3 times more mass there about 4600 million years BP compared to today (Jones, 1999).

Asteroids display a wide range of orbital characteristics including values for their semi-major axes, eccentricities and inclinations to the ecliptic (Encrenaz *et al.*, 1990). Most asteroids found in the main asteroid belt have a semi-major axis lying between 2 and 3.5 AU. Beyond 3.5 AU there are two distinct families of asteroids; *Hildas* at 4 AU and *Trojans* at 5 AU. There are also a number of small asteroids that have orbits that intersect with or closely approach the Earth's orbit (Taylor, 2001). These near-Earth asteroids are otherwise known as the *Apollo–Amor* family. They consist of asteroids originally derived from the main belt but which were ejected from the 'Kirkwood gaps' by Jupiter's gravitational perturbations (Encrenaz *et al.*, 1990; Taylor, 2001). There are a few thousand near-Earth asteroids that have diameters greater than 1 km. The *Apollo* asteroids are the current suppliers of meteorites to Earth (Taylor, 2001)

The size of asteroids is determined mainly through indirect measurements, with the exception of some of the larger asteroids such as *Ceres*, a few of the *Apollos*, and those that have been imaged from spacecraft enabling direct measurement (Jones, 1999). Most asteroids are too small and too far away to be seen as anything more than a point of light in the night sky (Jones, 1999). Indirect measurements are made, therefore, and this is achieved through the determination of the flux density of the reflected solar radiation received from that body (Jones, 1999). The flux density is the power of the electromagnetic radiation incident on a unit area of a receiving surface.

Asteroid types

The spectral characteristics of asteroids allow them to be grouped into four different classes; they can be further classified by their size and orbital path (Taylor 2001). C-type asteroids are carbonaceous and are the darkest asteroids. They are rich in hydrated silicates and in carbon. About 60% of known asteroids are of this type and are found in the outer portion of the main asteroid belt (Encrenaz *et al.* 1990). S-type asteroids are siliceous with spectral characteristics of rocky bodies that principally consist of pyroxene, olivine, as well as a metallic phase (iron and nickel). They represent 30% of catalogued asteroids and are abundant among the *Apollo–Amor* objects in the inner part of the main belt (Encrenaz *et al.*, 1990). M-type asteroids are metallic consisting entirely of iron and nickel (Encrenaz *et al.*, 1990). U-type are unclassified asteroids that do not fit this classification. The asteroid *Vesta* is an unclassified asteroid; it has a particularly high albedo and a spectrum dominated by pyroxene and feldspar absorption bands (Encrenaz *et al.*, 1990). Smaller asteroids are referred to as meteoroids and micrometeoroids which are then further classified to meteors and meteorites when they enter the Earth's atmosphere.

A meteoroid is a small asteroid less then 1 km across (Jones, 1999). Smaller particles are called micrometeoroids or cosmic dust grains, which includes any interstellar material less then 0.01 m across that should happen to enter our solar system (Encrenaz *et al.*, 1990; Jones, 1999). A meteorite is a meteoroid that reaches the surface of the Earth without being completely vaporised. Much of our understanding about asteroids comes from examining pieces of meteorites. Usually, meteoroids are too small though to survive the passage through the Earth's atmosphere. When a meteoroid strikes our atmosphere at high velocity, friction causes it to incinerate into a streak of light known as a meteor or 'shooting star' (NASA, 2004). The term meteor comes from the Greek word 'meteoron', meaning phenomenon in the sky. The temporary incandescence resulting from atmospheric friction of this matter typically occurs at heights of 80–110 km above Earth's surface.

Asteroid impacts with Earth

The damage caused by the impact of an asteroid with the Earth's surface depends upon the size of the asteroid, the landing location and the speed it

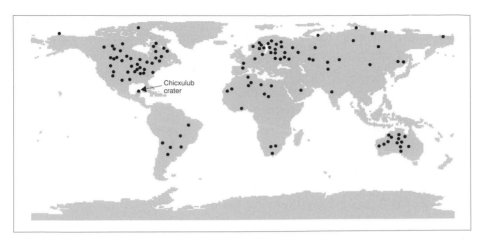

Figure 9.1. Distribution of craters globally. The Chicxulub (Yucatan Peninsula) crater is also shown (after Professor D. A. Kring, University of Arizona).

entered the Earth's atmosphere. The effects can range from damage to property as a result of a meteorite landing on a house or car, to the mass extinction of many species. The impact of asteroids can cause shock waves around the impact site, fill the atmosphere with dust and generate tsunamis capable of travelling whole oceans. To date no recorded human deaths from meteorite impacts have occurred. In more recent geologic time, there is evidence that at least one mass extinction event, notably that of the dinosaurs and many other species 65 million years ago, is linked to the global effects caused by a major asteroid impact. Impacts also have some economic significance; for example, the vast copper–nickel deposits at Sudbury, Canada are possibly related to a large-scale impact 1850 million years ago and several impact structures in sedimentary rocks have formed suitable reservoirs for economic oil and gas deposits (Earth Impact Database, 2003).

Stony meteorites with a diameter less than about 100 m generally do not reach the Earth's surface. These objects usually explode several kilometres above the surface and are known as an 'airburst' (NASA, 2004). The kinetic energy involved is substantial; a typical impact by a 50 m diameter object releases energy equivalent to about 10 megatons (Mt) of TNT and 100 m diameter object releases the equivalent of about 75 Mt. These are equivalent in energy to large thermonuclear explosions and could cause devastation over thousands of square kilometres. Larger asteroids that survive the journey through the Earth's atmosphere can have devastating effects and leave a definite signature on the Earth's surface and atmosphere.

Over 160 impact craters have been identified on Earth (Fig. 9.1). Almost all known craters have been recognised since AD 1950 and several new structures are found each year (Pesonen, 1996; Earth Impact Database, 2003). The basic types of impact structures on Earth are simple structures up to 4 km in diameter, with uplifted and overturned rim rocks surrounding a bowl-shaped depression, partially filled by breccia. Complex impact structures and basins are generally 4 km or more in diameter, with a distinct central uplift in the form of a peak and/or ring, an annular trough, and a slumped rim. The interiors of these structures are partially filled with breccia and rock melted by the impact (Earth Impact Database, 2003).

Impacts may also induce chemical changes in the atmosphere. These are related to the vaporisation of the impacting body and a portion of the target. Relatively small impacting bodies <0.5 km in diameter produce impact craters on the scale of 10 km in diameter and inject five times more sulphur into the stratosphere than occurs there at present. Larger impact events occurring on the timescale of a million years will inject enough sulphur to produce a drop in temperature of several degrees and a major climatic shift. There are additional effects on atmospheric chemistry, including the potential for the destruction of the ozone layer from shock heating atmospheric nitrogen and the injection of fluorides from the vaporised impacting body. The threshold for these effects appears to be on the time scale of 100 000s of years (Earth Impact Database, 2003).

In some circumstances an ocean impact might even be less hazardous to humankind than a land impact because less debris will be thrown into the atmosphere and indirect effects might be reduced. This was true for the 'Eltanin' impact, a 1–4 km diameter asteroid that landed in the ocean near Chile approximately 2 million years ago. It did not create a crater on the seabed and apparently did not result in mass extinctions (NASA, 2004). There is the risk though of destruction from asteroid-induced tsunamis when they strike oceans. For example, an impact anywhere in the Atlantic Ocean by a body 400 m in diameter would devastate the coasts on both sides of the ocean with wave run-ups potentially greater than 60 m. An impact-generated tsunami would be 10 times more powerful than the largest earthquake-generated tsunami and occur with a recurrence interval of a few thousand years.

The most devastating impact of an asteroid collision is a mass extinction. In 1980, Louis Alvarez and co-workers theorised that the Cretaceous–Tertiary (K–T) boundary sediments and elemental anomalies worldwide were due to a major asteroid impact 65 million years ago. There is an unexpectedly high concentration of the rare element iridium in geological samples taken from the narrow boundary between the Cretaceous and Tertiary periods and this time period also marks the disappearance of the dinosaurs from the fossil record (Zebrowski, 1997). One of the only other places where high concentrations of iridium are found is in metallic meteorites/asteroids. It was thought that a large metallic asteroid impacted with Earth emitting large quantities of iridium laden dust particles into the upper atmosphere. Eventually this dust settled uniformly over the planet. Such an event could depress temperatures and the amount of surface-reaching sunlight around the globe, leading to loss of food crops and related problems over a number of years, creating inhospitable conditions for the survival of dinosaurs (Zebrowski, 1997).

The impact signal of the K–T event is recognisable globally (Smit, 1999). Large impact events have the capacity to blow a hole in the atmosphere above the impact site, permitting some impact materials to be dispersed globally by the impact fireball, which rises above the lower atmosphere (Taylor, 2001). These materials do not require atmospheric winds for dispersal and have the capacity to encircle the globe in a relatively short time, before eventually returning to the surface. Relatively small impact events, resulting in impact structures in the 20 km size range, can also produce atmospheric blow-out (Jones, 1999). However, at present the K–T is the only biostratigraphic boundary with a clear signal of the involvement of a large-scale impact event. The involvement of impacts in other geological boundary events in the terrestrial stratigraphic record has been suggested but little evidence has been offered.

The risk of an asteroid impact

Asteroids and comets have collided with the Earth throughout its 4.5 billion year history. The impact of near-Earth objects with our planet can be catastrophic and still represents a natural hazard today (Wetherill and Shoe-maker, 1982). The most dangerous asteroids are those capable of causing major regional or global disasters; fortunately these are extremely rare. These bodies impact the Earth only once every 100 000 years on average (NASA. 2004). The greatest risk is associated with objects larger than 1–2 km diameter, which are large enough to perturb the Earth's climate on a global scale. From estimates of the terrestrial cratering rate, the frequency of the Cretaceous–Tertiary-sized events occurring is about one every 50–100 million years (Earth Impact Database, 2003). Meteorites of about 0.5 km diameter impact the Earth once every 10 000 years and produce craters about 10 km in diameter (Wetherill and Shoemaker, 1982).

The formation of impact craters as large as 20 km could produce light reductions and temperature disruptions similar to a nuclear winter. Such impacts occur on Earth around once every 2–3 million years (Earth Impact Database, 2003). The most recent known structure in this size range is 'Zhamanshin' in Kazakhstan, with a diameter of 15 km and an age of 1 million years BP. Impacts of this scale are not likely to have a serious effect upon the biosphere and cause mass extinctions (Wetherill and Shoemaker, 1982). Events such as Tunguska, an airburst about 8 km above Siberia, occur on a timescale of hundreds of years.

Impacts with Earth are thought to occur randomly in both space and time. The next large impact with the Earth could be an 'impact-winter' producing event or even a K–T-sized event (Earth Impact Database, 2003). To emphasise this point, in March 1989 an asteroidal body named 1989 FC passed within 640 000 km of the Earth. This Earth-crossing body was not discovered until it had passed the Earth (NASA, 2004). It was estimated to be in the 0.5 km size-range, weighing 50 million tons and travelling at 74 000 km h^{-1}. An object of this size is capable of producing a Zhamanshin-sized crater or a devastating tsunami; however, it missed the Earth by only 6 hours (Earth Impact Database, 2003). At present, no systems or procedures are in place for mitigating the effects of an asteroid impact. In February 2004, the 'Protecting the Earth from Asteroids' conference in the United States concluded that the world's governments were not prepared for a short-term threat from a potential asteroid impact. The conference recommended that future resources be channelled to developing response strategies for an imminent impact and that all avenues to dislodge an asteroid from its collision course (including nuclear weapons) should be pursued (NASA, 2004).

The asteroid, known as 1997 XF11, will pass within the Moon's distance of the Earth in October 2028 (NASA, 2004). This distance may seem large in human terms, but it is less than has previously been predicted in advance for any other known asteroid during the foreseeable future.

Historical events

The most famous asteroid impact was 65 million years ago, when about 70% of all species then living on Earth disappeared within a very short period (Figs. 9.1 and 9.2). The K–T impact occurred in the Yucatan region of Mexico. It is thought that the asteroid was about 10 km wide. Chicxulub, the name given to the asteroid, impacted at a velocity of 11 km s^{-1} creating a 180 km diameter ring structure (Zebrowski, 1997). The event threw huge amounts of matter into the atmosphere and generated 600 m high waves that may have emptied the Gulf of Mexico (Zebrowski, 1997). It is thought that the impact created months of darkness and much cooler temperatures globally (Cockell, 1999). The relic impact crater formed by Chicxulub is centred on the present coastline of Yucatan and now lies beneath several hundred metres of sedimentary deposits. It was only discovered in 1990 from geophysical data taken 10 years earlier while searching for oil (Zebrowski, 1997; Earth Impact Database, 2003).

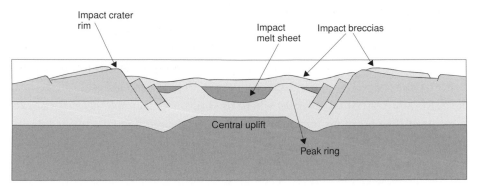

Figure 9.2. Schematic of the Chicxulub impact crater (after Professor D. A. Kring, University of Arizona).

The biggest asteroid impact on Earth to have occurred in modern times is known as the Tunguska event. On the 30th June 1908 a small asteroid or meteroid, 100 m in diameter, exploded 8 km above the remote region of Tunguska in Siberia. The explosion produced a shock wave that knocked people to the ground in the town of Vahaurara 70 km away (Vasilyer, 1998). The object was reduced to dust by the explosion and stratospheric winds spread the dust around the globe. Seismic records of the event recorded it as having an explosive energy of 12.5 million tons of TNT. The explosion destroyed about 2150 km^2 of Siberian forests felling some 60 million trees (Vasilyer, 1998). The event inspired a plethora of scientific investigations and debate over the last century as to the cause of the event but the evidence to date suggests that it was an asteroid, not a comet as first suggested, that vaporised as it entered the Earth's atmosphere.

On the 9th October 1992 a meteorite (H6) of 12.4 kg landed on a parked car in Peekskill, USA. As the meteorite entered the Earth's atmosphere it created a fireball that was reported to be as bright as a full moon and was seen streaking across the sky from Kentucky to New York. At least 14 people captured part of the fireball on videotape. On the evening of the 26th March 2003 meteorites punched through roofs in Park Forest, Illinois, USA. It was believed that the meteorites came from a larger mass that weighed no less than 900 kg before it hit the atmosphere (Earth Impact Database, 2003).

PALAEOASTEROID IMPACTS WITH EARTH

Despite the fact that approximately 160 impact craters have been identified on Earth, such features are difficult to detect because terrestrial processes such as volcanism, tectonics and erosion can erase the impact record. The presence of meteoritic components in geological strata is important in the identification of impact events. Shock metamorphic effects on various minerals such as quartz and zircon are some of the most reliable features used to identify extraterrestrial impact events. These features are regarded as reliable indicators of past impact events because shock features in minerals related to non-impact processes have never been confirmed in nature. Impact-derived rocks (impactites) form from terrestrial target rocks where the impactor is completely vaporised as a result of the enormous temperatures that are released during a hyper-velocity impact. Minor amounts of recondensed meteoritic vapour may be mixed with the vaporised, molten or shocked and brecciated target rocks and such debris can be spread over vast distances across the Earth's surface. These processes are only applicable for craters greater than 1.5 km diameter. For smaller events, only small parts of the impactor may be preserved and here, isotopic signatures in suspected impacted rocks and ejecta material are used to fingerprint an extraterrestrial source. Most research on extraterrestrial impacts, therefore, is based on stratigraphic and geochemical analysis, especially where the fossil record shows major extinction events have occurred.

Impact craters: processes and effects

Craters are the most obvious features marking the impact of prehistoric bolides with the Earth and other planets. However, many features can be mistaken for impact craters because, upon initial inspection, they can exhibit a similar morphological expression. Features such as solution depressions in both limestone and other types of rock can resemble impact craters because of their general circular shape and dish-like depression. Impact craters do have distinctive features though that separate them from these other landforms. On Earth, weathering and erosion can remove the crater-like morphology relatively quickly compared to other planetary bodies such as the Moon where tectonic activity is absent. Where preserved, these features are termed impact structures or astroblemes.

Impact cratering can undergo three stages of formation. These are the initial contact and compression, excavation and formation of a transient crater and allochthonous breccia (rock fragments detached and displaced from their original position), and modification with slumping and readjustment (central uplift and annular collapse, Sturkell, 1998) (Fig. 9.2). Rondot (1994) describes several processes and features directly linked to the impact and formation of impact craters. These are detailed below.

1.   *Effects of pressure on the rocks*

At impact, the kinetic energy of the bolide (any natural object colliding with Earth) is transferred to the target area resulting in compression. The shock pressure reaches hundreds of GigaPascals (GPa) and the temperature rises thousands of degrees. The increased pressure results in rocks diminishing (volume, thickness) as they exceed their Hugoniot elastic limit (HEL). The HEL is generally below 1 GPa for sedimentary rocks and between 3 and 4.5 GPa for crystalline rocks. At higher pressures, when the volume of rock diminishes more slowly than it does immediately after the HEL is exceeded, the two denser minerals stishovite and coesite may form at 15 and 25 GPa, respectively. At pressures above 40–50 GPa, these minerals disappear to form silica glass. At even higher pressures, the whole rock melts and vaporises.

2.   *Effects of the shock wave on minerals in crystalline rocks*

The peak pressure during an impact occurs during a fraction of a second. The effects of the shock wave in minerals are complex as rock is typically heterogenous and usually contains several minerals that possess different shock reactions. For example, quartz develops planar deformation features during shock impact. Different types of planar features (four in total) have been identified; their occurrence depends upon the pressure during impact.

3.   *Distribution of shock metamorphosed rocks*

At impact, the shock wave continues to propagate in the impact area taking the shape of a hemisphere centred roughly one projectile diameter below the ground surface. As the shock wave expands, it degrades into a plastic stress wave that is preceded by an elastic precursor wave before it changes to an entirely elastic wave below 5 GPa. The reaction of the rock to these waves depends upon the strength of that rock.

4.   *Fine, regularly spaced fractures, shatter cones*

Impact craters are usually characterised by an abundance of fractures. This is not a direct result of the shock wave, but of tensile interactions with subsequent refraction waves.

5.   *A-type pseudotachylite and B-type pseudotachylite*

Thin veins of pseudotachylite form almost without displacement in the centre of astroblemes. A-type pseudotachylite forms as a result of local fusion of the rock material. B-type pseudotachylites are thick autochthonous breccia dykes. They are highly varied in nature and can be white, grey, black, green or red when altered by weathering near the surface.

6.   *Impactite dyke or impact melt dykes*

These also characterise large astroblemes and consist of dykes of fine crystalline rock, locally, with fragments of shocked rock and quench textures.

These dykes are considered to be impact melts that are driven into the expanding cavity as high-velocity turbulent flows of superheated low-viscosity silicate liquid that penetrate downwards into fissures and cracks at the bottom of the crater.

Despite the variety of forms of evidence that can be used to identify impact structures it can still be difficult to identify these features when they are of great age and have been subject to extensive erosion and post-impact deformations. The three main diagnostic criteria for identification (historical record of an impact event; meteorite fragments or contaminated meteorite material in rocks such as enriched siderophile element abundances; and shock metamorphic features such as shatter cones or planar deformation features or solid state or fusion glasses) may not always be immediately obvious. In such cases, other criteria such as circular gravity and magnetic anomalies, impact-produced igneous melts or breccia dykes, circular geochemical anomalies, and linkage between distal ejecta layers to specific structures may need to be found.

Barnouin-Jha and Schultz (1998) discovered that ejecta lobateness resulting from instabilities created in an atmospheric vortex ring are a useful diagnostic criterion. The atmospheric instabilities or waves are generated by the advancing ejecta curtain and the number of ejecta lobes on the ground increases with the size of the advancing vortex size scaled to its core radius and the square root of the Reynolds number (description of the degree of turbulence) of the flow in the vortex ring. The vortex ring forms behind the impermeable portion of the ejecta curtain, as it passes through the atmosphere, by flow separation and rapidly begins to entrain ejecta out of the curtain. As the distance between the vortex ring and the curtain increases, the vortex ring becomes unstable, creating waves. These waves force the vortex to evolve from a laminar to a turbulent regime (as described by the Reynolds number). The vortex ring then drops towards the target surface before it continues to scour and entrain ejecta as it advances along the ground surface. Finally, the vortex ring disintegrates into multiple lobes of ground-hugging flows of ejecta debris resembling a turbidity flow. There is good correlation between observations of ejecta lobateness and theoretical calculations, and observations of lobes on Mars and Venus.

Two of the main factors that play a part in the preservation of prehistoric impact craters and events are the magnitude of the event and the degree of consequent deformation, recrystallisation and melting. Impact models for asteroids with diameters larger than 10 km, especially where they impacted on thin and hot Archaean (the most ancient of Earth's rocks) crust, suggest that the thermal and magmatic effects of a large magnitude impact can be expected to obliterate the proximal impact deformation signals through recrystallisation, shearing and melting. This means that shock deformations and melting

Figure 9.3. The K–T boundary sequence at Raton Basin in Colorado. The K–T boundary is marked by the light coloured unit (knife). Photograph courtesy of Professor D. A. Kring, University of Arizona.

effects may not be recorded stratigraphically and lithologically. However, distal impact effects, often consisting of deposition of microtektites and diamictites, could be expected to be preserved in the stratigraphic record (Glikson, 1993).

Impact events can also generate major oil-bearing rock reservoirs. Grajales-Nishimura *et al.* (2000) analysed the lithology, biostratigraphy and mineral composition from the offshore wells on the sea bed near the K–T impact event (Fig. 9.3). They found a unique stratigraphy and distribution of impact material within the calcareous breccia of the marine platform which they assume was generated during the impact event. Immediately following the impact, the offshore carbonate platform collapsed resulting in the deposition of a breccia unit. Following this, ballistic impact ejecta was deposited and then much of this material experienced reworking and mixing with coarser material as impact-generated tsunamis were reflected back and forth in the Gulf of Mexico. Grajales-Nishimura *et al.* (2000) suggest that these extraterrestrial impact-generated structures, seals and reservoir facies are conducive for oil production. The K–T breccia generated during this event is now one of the best known oil-reservoir rock units related to an impact event.

Figure 9.4. Shocked and unshocked quartz grains. The unshocked grain is to the left. Note the striations (shock marks) in the grain on the right. Photograph courtesy of Professor D. A. Kring, University of Arizona.

SHOCK PROCESSES IN QUARTZ AS A DIAGNOSTIC TOOL

Quartz is a relatively stable rock-forming mineral that occurs nearly ubiquitously in terrestrial crustal rocks. It does not normally display shocked textures and, where it does, it can be regarded as a reliable indicator of extraterrestrial impacts (Fig. 9.4). Shocked quartz is found in only two geological settings: either in direct spatial relationship to impact structures or as a constituent of distal or global ejecta deposits. Shock effects vary with increasing shock pressure and also to some degree on the pre-shock temperature. Quartz starts to melt under shock compression of >50 GPa and vaporises at >100 GPa. Grieves *et al.* (1996) note that between about 5 and >50 GPa the following shock effects in quartz are observed.

(1) Mosaicism which is a highly irregular mottled extinction pattern (as seen under a polarising microscope).
(2) Planar microstructures (planar fractures and planar deformation features). There are two ways in which quartz fails mechanically. These are by the formation of irregular fractures that are not diagnostic shock effects and by the formation of regular planar microstructures that are diagnostic shock effects. There are two types of planar microstructures: planar fractures (PFs) and planar deformation features (PDFs).
(3) Partial to complete isotropization or amorphisation (diaplectic glass). Diaplectic glass has the same morphology as the original quartz mineral and shows no evidence of fluid textures. The refractive index of diaplectic glass decreases with increasing pressure until shock-fused quartz (lechatelierite) is formed.
(4) Partial transformation to stishovite and coesite. These are high-pressure polymorphs that occur as fine-grained aggregates formed by partial

transformation of the host quartz. The crystallisation and textural set-ting of these phases are different in crystalline rocks and in porous sedimentary rocks. Coesite is formed between 30 and 60 Gpa, while stishovite is formed between 12 and 45 GPa.

(5) Melting and quenching to form lechatelierite. This is a product of the highest degree of shock and is commonly found at impact craters formed in sedimentary rocks or unconsolidated sediments as highly vesicu-lated glass with flow structures. In non-porous, crystalline target rocks, lechatelierite is only found in impact melt glasses and rocks in the form of inclusions and schlieren that results from secondary melting of highly shocked quartz mixed into the impact melt during the crater-forming process.

Simplified shock wave profiles in quartz are characterised by three phases: the loading phase, the compression phase and the unloading or decompression phase. Complexities in the shock wave structure, however, can result in phase transitions. Three temperatures are associated with shock metamorphism, these being the pre-shock, the shock and the post-shock temperature. The latter two are a function mainly of the shock pressure. Shock temperatures in porous rock tend to be significantly higher than in crystalline rocks. Distal ejecta are ejecta that are well removed from the impact crater, highly dispersed and quickly cooled. Quartz and PDFs have been discovered in distal ejecta deposits at the K–T boundary (Grieves *et al.*, 1996).

Impact ejecta and spherules

Extraterrestrial impacts can generate a layer of debris mantling the area surrounding the crater (the ejecta blanket) to more distal ejecta fallout lay-ers. These deposits may be preserved in the geological record and range from a local to global extent. The energy released by extraterrestrial impacts can also directly or indirectly trigger a variety of high energy, surficial processes that can rework sediments surrounding the impact site. These include wave sys-tems, subaerial or subaqueous sediment gravity flows and possibly hyper storms. The deposits generated by these catastrophic events are likely to be preserved in the sedimentary record, particularly in otherwise low-energy environments (Hassler and Simonson, 2001). The extent of the distribution of the impact ejecta is related to the impact energy, and therefore also the final diameter of the impact structure. To distribute impact ejecta globally, impact-shattered and molten target rocks have to be ejected past Earth's atmosphere. This is known as an atmospheric blow-out. The ejecta is further distributed by atmo-spheric winds as it returns to the atmosphere by gravitation. Impact diameters as small as 3 km are capable of producing atmospheric blow-outs (Whitehead *et al.*, 2000)

Spherules are small fragments of melted rock blasted into the atmosphere after bolides impact with the ground surface. They are deposited into thin layers at various locations across the Earth's surface. They occur across a variety of geological time periods but appear to be most common within Precambrian strata of South Africa and Western Australia. The Precambrian spherule layers were deposited between 3.46–3.23 and 2.64–2.49 Ga (billion years ago). Strata from the Phanerozoic, the geological time period following the Precambrian, contains fewer spherule layers. The most famous of these, however, are the thin layers associated with the Cretaceous–Tertiary (K–T) boundary impact event. Simonson and Harnik (2000) recognised that the K–T boundary layer is usually only a few millimetres thick but some deposits, thousands of kilometres away from the Chicxulub crater, can be 60–100 mm thick.

Tektites are glass bodies, lacking internal crystals, formed by the melting of terrestrial surface deposits during extraterrestrial impacts. They are found in four major strewn fields, these being the Australasian, Ivory Coast, Czechoslovakian and North American fields. Tektite layers are associated with unmelted impact ejecta in the form of shocked quartz, feldspar, coesite and stishovite in both the North American and Australasian strewn fields. The source crater for the tektites and microtektites in the Australasian strewn field is in the Indochina region (Glass and Wu, 1993) (Fig. 9.5). Coesite and shocked quartz are absent from the Ivory Coast strewn field. This may be due to a smaller-size event and the distance of the sample sites from the impact area.

Microkrystites consist of a combination of crystals (primarily clinopyroxene) and glass (usually replaced) (Simonson and Harnik, 2000). Deep-sea microspherules consisting of microtektites and microkrystites are also found globally in upper Eocene rocks, around 40 million years ago, and are generally believed to have been generated by an extraterrestrial impact melting and ejecting terrestrial material (Wei, 1995). The number and exact age of these layers varies between locations.

The Hammersley Basin of Western Australia has four spherule layers, which formed from extraterrestrial impacts between 2.49 and 2.63 Ga (Hassler and Simonson, 2001). It would appear that these spherule layers were deposited below the wave base in a basinal or a deep-shelf marine environment because they are:

(1)   interbedded with sedimentary rocks that were originally clastic or chemical mud, including carbonate, shale, chert, and/or banded iron formations;

Figure 9.5. Tektites. Photograph courtesy of Professor D. A. Kring, University of Arizona.

(2) ubiquitous in the muddy strata, some with extreme lateral continuity; this indicates that they were deposited in quiet water and not influenced by shallow-water processes; or

(3) found close to rare, coarse-grained strata also indicating deep-water deposition.

The spherules consist mainly of K-feldspar crystals that are organised into radial-fibrous sprays that diverge inward from the grain edges. These textures are similar to devitrification textures from both lunar impact spherules and spherules in the K–T boundary layer. Spherules with random textures were also found. These random textures are interpreted to be a product of primary crystallisation during the cooling of droplets that were originally basaltic in composition. Replacement of K-feldspar, such as occurred in some of the central spots of the radial textures, suggests a basaltic composition. The spherule layers are distinguishable from volcanic deposits, as the volcanic ash at these locations is fine grained and massive or displays a crude concentric zoning. The spherules are also distinguishable from carbonate ooids. Iridium and other platinum-group element anomalies are also present in the layers as well as other particles exhibiting shapes associated with ejecta and irregularly shaped melt-ejecta particles. Planar deformation features are not evident in the stratigraphy of these sites suggesting that the target area was the ocean floor (Hassler and Simonson, 2001).

All four spherule layers in the Hammersley Basin sequence appear to have been affected by substrate erosion, wave activity, offshore-directed bottom flows and sediment gravity flows. These processes are unusual for deep shelf, quiet water environments which further suggests that the spherule layers resulted from extraterrestrial impacts. The most common explanation for generation of large waves in such circumstances is the collapse of the transient water crater generated during an oceanic impact, leading to the formation of tsunamis. However, impact-triggered seismic activity leading to submarine landsliding and water displacement by ejecta may also have been responsible. It is not possible to determine which of these processes triggered the large wave action, because there is a lack of modern analogues of impact-generated tsunamis.

Impact ejecta is found globally in outcrops and drill cores at the K–T boundary (Smith and Ward, 1998). Although not necessarily universally accepted, the K–T boundary impact ejecta is widely believed to have been derived from the Chicxulub impact crater, Mexico. Evidence of a mass extinction at the K–T boundary is still, however, somewhat controversial. The most widely held view is that a mass extinction event, including the demise of the dinosaurs, immediately followed deposition of the K–T boundary layer sequence. Alternative views suggest that a gradual change in the composition of flora and fauna can be found in the upper Cretaceous record, indicating that the extinctions were the result of a deteriorating climate. However, it might be difficult to distinguish between a mass extinction resulting from climate deterioration due to an extraterrestrial impact and climate change from other causes. There is little doubt, however, that the K–T boundary layer was deposited following a large asteroid impact. The K–T boundary layer is associated with iridium anomalies, shocked minerals and impact spherules. Four types of ejecta deposits can be distinguished at the K–T boundary. Smith and Ward (1998) state that these are:

(1)   an ejecta layer, a few millimetres thick, which is found globally;
(2)   one to two centimetre thick ejecta layers are visible in the stratigraphic record of the western interior of the USA;
(3)   high-energy (tsunami) clastic deposits are found in Cretaceous to earliest Cenozoic rocks in the Gulf of Mexico region; and
(4)   ejecta blanket deposits are found up to 3.5 crater radii away from the crater rim (Gulf of Mexico).

Volcanism has been suggested as an alternative cause of the Cretaceous–Tertiary extinctions. However, Smit and Kyte (1985) suggest that it is unlikely that a volcanic event is capable of a worldwide distribution of spherules as well as numerous examples of shocked quartz such as that found in K–T boundary sediments. Although volcanism cannot be excluded as a possible source of these features, Smit and Kyte (1985) state that it is statistically more probable for an impact to be a more likely explanation than a poorly defined mantle event resulting in worldwide simultaneous volcanic eruptions of unprecedented magnitude.

Spherule layers have been found at more distant locations from the proposed Gulf of Mexico K–T impact site. Norris *et al.* (1999) examined cores from the northwestern Atlantic region and described the biostratigraphy and lithology for the K–T boundary. This time zone was marked by a 10 cm thick spherule layer assumed to have been derived from an extraterrestrial impact event. The composition of clasts within the spherule bed suggests that the grains were derived from either a metamorphic basement or a continental source area associated with a carbonate platform. This is consistent with the subsurface geology in the area around the Chicxulub crater. The size of the spherules also correlates well with spherules identified in the Gulf of Mexico. An Ir anomaly in the upper part of the spherule bed is interpreted as further evidence of deposition of an ejecta layer following an extraterrestrial impact (Norris *et al.*, 1999).

Spherule beds are not always attributed to impact events. Koeberl *et al.* (1993) examined several such beds in the Archean Barberton Greenstone Belt, South Africa (Reimold *et al.*, 1997). While previous studies had suggested extraterrestrial impacts as the cause of the spherule beds (Lowe and Byerly, 1986), Koeberl *et al.* (1993) suggest that the mineralogy of the sequence indicates that extensive hydrothermal activity of volcanic rocks is a more likely cause. They also note that the spherule layers consist of several distinct sublayers, which if the impact event hypothesis were proposed, would mean that a number of separate events would have needed to have occurred over a 30 Ma interval at the one site. Deposition of the spherule layers at Barberton are dated to about 3.2 Ga, and while others have suggested that the multiple event hypothesis could have been due to the period of the Late Heavy Bombardment, Koeberl *et al.* (1993) note that this ended about 3.8 Ga, hence before deposition of the Barberton spherules. It is also unlikely that three distinct impacts could occur in the same region especially when this number of spherule layers from this time interval has not been recognised elsewhere. Koeberl *et al.* (1993) further noted that the microtektites found here occur in the form of solid beds, which is unusual for well documented impact sites. Such microtektites are also much smaller than the Barberton spherules, and are dispersed over a relatively large area as well as having petrological and chemical characteristics that differ from those of the Barberton spherules. Koeberl *et al.* (1993) suggest that a better explanation is that the spherule layers have resulted from widespread volcanic activity, followed by several phases of hydrothermal and tectonothermal activity leading to siderophile element mobilisation and redistribution in the course of sulphide mineralisation.

Simonson and Harnik (2000) found that there appear to be differences between Precambrian and Phanerozoic spherule layers. On average, early

Precambrian spherule deposits are found in thicker accumulations and are more crystallised. These spherules appear to have formed from basaltic melts rather than from continental target rocks as is the case with spherule layers from the Phanerozoic. No glass or original minerals have been found in early Precambrian layers. Simonson and Harnik (2000) suggest that while these differences could indicate that there have been secular changes to variations in the nature of impact ejecta, changes in the nature of Earth's hydrosphere and lithosphere might be an equally important reason. As most of the mass in distal melt and condensate particles from impacts comes from the target crust, it is likely that changes through time in the average composition of spherules probably reflect changes in the average composition of crustal rocks. It is likely that the volume of continental crust was significantly smaller in Archean times and the ocean crust proportionately larger. This would have resulted in a greater percentage of mafic rocks in the target material and thus likely to be recorded in the distal impact ejecta. More open-ocean impacts could also skew the nature of distal ejecta in stratigraphic layers towards thicker accumulations. Another important fact to consider is that only the largest impacts, where the diameter of the projectile is approximately the same as the water depth, excavate and disperse target rocks from the sea floor; the water mass shields the target area in the case of smaller impacts. A predominance of mainly mafic target rocks could also explain the lack of shocked quartz in early Precambrian spherule layers.

Spinel

Spinel is a mineral frequently found in igneous rocks. However, when it is high in nickel (Ni) and is oxidised, it is more likely to have formed during hypervelocity meteoritic interactions with the Earth's atmosphere. Meteorites have relatively high Ni content along with a high ferric/ferrous ratio resulting from crystallisation in the O_2-rich environment of the Earth's atmosphere. Pierrard et al. (1998) note that spinel is found in meteorite fusion crusts, cosmic spherules from deep sea sediments, polar ice and a variety of meteoritic debris in the sedimentary record. Oxidised Ni-rich spinel does not form by volcanic processes as terrestrial magmas are depleted in nickel and evolve under extremely low oxygen fugacity environments (Robin et al., 1992). Spinel is found at a number of distinct time periods in the geological record such as the K–T boundary, the lower to middle Jurassic in the southern Alps of Italy, late Eocene sediments in Italy, late Oligocene sediments (~30 Ma) from the central north Pacific and in late Pliocene (~2 Ma) sediments in Antarctica. The presence of spinel and the size of spherules can also potentially provide constraints on the size of the impact, the nature of the early Archean atmosphere and transient effects produced on it from the impact and nature of the bolide. Together with Ir anomalies, spinels with high Ni content are often the only preserved evidence of distal impact deposits.

Robin *et al.* (1992) have classified spinel occurrences into three groups according to their chemical compositions.

(1)   A micrometeorite spinel formed by oxidation of cosmic dust particles in the upper part of the atmosphere, at relatively low oxygen fugacities. It is mostly magnetite characterised by an Fe^{3+}/Fe_{total} ratio up to 75–80 atom%.

(2)   A meteorite spinel formed by ablation of meteorites in the lower part of the atmosphere, at higher oxygen fugacities. It is a solid solution of magnetite and magnesioferrite characterised by an Fe^{3+}/Fe_{total} ratio between 75 and 90 atom%.

(3)   A meteorite spinel formed by interaction of impact debris with the highly compressed atmosphere. It is essentially magnesioferrite characterised by an iron oxidation state ($Fe^{+3}/Fe_{total} > 90$ atom%).

The Barberton Greenstone Belt of Southern Africa contains four impact layers (which is disputed by Koeberl *et al.*, 1993) and two of these layers contain Ni spinel and Ir anomalies (Byerly and Lowe, 1994). Of the other layers, one exhibits only Ir anomalies and the other none. However, all layers are characterised by the presence of spherules. Despite the debate concerning the extraterrestrial origin of these beds in South Africa, the presence of Ni-rich oxidised spinel elsewhere around the globe as a signature of bolide impacts is not disputed.

One of the most widely observed spinel-bearing geological units (often multiple microtektite–microkrystite layers) occurs at the boundary between the Eocene and Oligocene time epochs around 35 Ma. Pierrard *et al.* (1998) examined the global type section for the Eocene–Oligocene boundary in Massignano, Italy where they aimed to determine whether Ni-rich spinel is present. Pierrard *et al.* (1998), using the Robin *et al.* (1992) chemical classification, established that the spinel from the Massignano site belongs to the first described class as it consists mainly of magnetite and has an Fe^{3+}/Fe_{total} ratio between 70 and 80 atom%. It is possible, therefore, that the spinel layer found in the upper Eocene could have resulted from a comet shower, although Pierrard *et al.* (1998) suggest that this is not likely as the magnitude and high accretion rate of spinel in this horizon indicate a sudden event. A comet shower is likely to span over a period of a few million years. It is more likely that the spinel layer originated from a large cometary impact. Because dust particles are imperative in the formation of the micrometeorite spinels in group 1, Pierrard *et al.* (1998) lean toward a comet impact rather than an asteroid as comets consist mainly of ice and dust. Spinel associated with the K–T impact, however, suggests it was formed by ablation of a large object in the lower atmosphere suggesting that this was an asteroid rather than a comet.

Iridium and other platinum-group elements (PGE) as indicators of extraterrestrial impacts

Iridium is one of the platinum-group elements. Iridium concentrations are very low in the Earth's crust, and high iridium values in crustal rocks can be assumed to be an indicator of a bolide impact. The most famous iridium-rich anomaly is associated with the Cretaceous–Tertiary boundary but iridium-rich anomalies have also been associated with other boundaries in the stratigraphic record. Other element anomalies have also been detected in the platinum-group elements. The platinum-group elements can be divided into two groups, the Ir group (Iridium, Ir; Ruthenium, Ru; Osmium, Os) and the Pt group (Platinum, Pt; Rhodium, Rh; Palladium, Pd; Gold, Au). There is a continual input of Ir and other PGE from micrometeorites as a relatively constant background accumulation rate occurs in marine sediments. However, at times the stratigraphic record shows that this background is modified periodically by large asteroids and comets. The anomalies assumed to arise from impacts of asteroids and comets are particularly evident in deep-sea sediments, and may be detected worldwide. Extraterrestrial material may not only supply Ir to seawater in detrital form, but also as a result of vaporisation of dust in the atmosphere and shock vaporisation of impacting large asteroids. Material enriched in Ir and other PGE does not only originate from the asteroid or comet itself, but from the impact ejecta. Whereas oceanic impacts result in relatively 'pure' extraterrestrial material, terrestrial impacts include both extraterrestrial and target material in the impact ejecta from the centre of the developing crater. Thus, in these cases, the abundance of Ir and other PGE depends not only on the asteroid composition, but also on the type of target rock and the relative masses of asteroid and ejecta materials (Sawlowicz, 1993).

Although it has been assumed that iridium anomalies are related to asteroid impacts, it has been suggested that natural processes also have the potential to cause similar anomalies. Volcanism in particular is a major process of Ir enrichment. Enormous amounts of basalt, known as the Deccan Traps, were erupted throughout western India around the time of the K–T boundary. Although such events can possibly supply enough Ir to cause a marked enrichment, the content of Ir in the Deccan Basalts is low, and it is unlikely, therefore, to have contributed sufficient Ir to produce the widespread increase in Ir at this time. Not all volcanic products, however, are necessarily low in Ir. Ir can be released from hot spot volcanoes as gaseous fluoride compounds, and it can be released as PGE carbonyls within volcanic aerosols. High Ir concentrations have been found in ashes from the Mt Kilauea volcano, Hawaii. These iridium-rich particles are likely to be derived from deep magma sources and it is possible that they can be adsorbed onto organic and inorganic particles. Sawlowics (1993) suggests that deposition of these particles, or dissolution in seawater, can possibly explain some enhanced Ir and PGE levels. Also, because there are few PGE data sets for mantle-derived rocks, it is difficult to differentiate between anomalies derived from terrestrial and extraterrestrial sources (Evanc *et al.*, 1993).

Biological processes also appear to be able to concentrate Ir from seawater. Some bacteria and fungi can dissolve Ir from both igneous rocks and from meteoritic materials. The mobilised elements can then be concentrated in the sedimentary environment by living or dead cyanobacterial mats, other bacteria and fungi. Micro-organisms can also dissolve Ir during diagenesis and redeposit it in adjacent sediments (Sawlowics, 1993). Soils, marine and freshwater organisms, therefore, appear able to dissolve and/or concentrate Ir and other PGE. Dyer *et al.* (1989) demonstrated that some micro-organisms can enhance, disperse or erase catastrophic iridium anomalies. They can also create anomalies from terrestrial sources of iridium. Dyer *et al.* (1989) showed that through the redistribution of iridium over short distances in sediments, micro-organisms may be responsible for the presence or absence of iridium anomalies as well as for the apparent time interval represented by an anomaly. Common micro-organisms worldwide appear able to affect the chemistry of iridium over a wide pH range by concentration or dissolution. It is important, therefore, that the microbial chemistry is taken into account when examining Ir anomalies associated with extraterrestrial impacts as conclusions about the nature of an event, the size of the impact body and the scale of an event may be misinterpreted (Dyer *et al.*, 1989). Despite the role that biological processes may have in producing Ir anomalies, there is little doubt that such anomalies, when all other factors are taken into account, are a signature marker for past bolide impacts with Earth.

Along with the K–T boundary, Ir anomalies occur during the Cambrian, Devonian and Tertiary, some of which also correspond to major extinction events. Wallace *et al.* (1991) focused on analysing the sedimentary Ir anomalies found near the Frasnian–Famennian (F–F) boundary (Upper Devonian) in the Canning Basin, Western Australia. This is one of the major extinction event boundaries. The mass extinction event appears to have occurred slightly after the Ir anomaly in the stratigraphic record at this time. Although meteorite impact has been suggested to explain the Ir anomaly and the mass extinction events, it has also been suggested that the filamentous microbe *Frutexites* may have collected Ir and also Pt directly from seawater. The Ir anomaly is found together with this microbe in a ferruginous stromatolite bed in the stratigraphic record within the Canning Basin. Wallace *et al.* (1991) state that several factors suggest that the Ir rich layers here, and other sites with similar characteristics, are of terrestrial origin because:

(1) no shock-metamorphosed minerals have been detected at the F–F boundary;

(2) the geochemistry of ferruginous crusts is invariably non-chondritic (both platinum metals and other trace elements);

(3) downward-orientated ferruginous crusts in cavities have similar platinum metal concentrations; this suggests a solution source (seawater) rather than particulate ejecta products; and

(4) association between ferruginous stromatolitic horizons and subareal exposure events suggest that changing sea levels (regression followed by transgression) have indirectly controlled the crust formation.

As Cambrian, Devonian and Tertiary horizons investigated in the Canning Basin show evidence of subaerial exposure in shallow-water lithologies, but not in deep-water sequences, it is assumed that the regressive events were of moderate scale. It is also possible that the ferruginous crusts are more directly related to rapid sea-level rise than regression. Further research is needed, however, to determine this. Together, these factors indicate that the Ir anomaly found at the F–F boundary is related to microbial processes and sea-level change (most likely rapid drowning). It is also possible that inorganic processes such as surface adsorption of platinum metal complexes onto positively charged Fe-hydroxide particles created the anomaly (Wallace *et al.*, 1991).

Nicoll and Playford (1993) also examined the Ir anomalies supposedly found at the F–F boundary in the Canning Basin, Western Australia. They placed more specific time constraints on the anomalies and were able to show that Ir anomalies are located significantly both above and below the F–F boundary and the major extinction event. They also found no direct evidence of a major extraterrestrial impact.

The Hagenberg event, a mass extinction event found at or near the Devonian–Carboniferous (D–C) boundary, has also been associated with an Ir anomaly. Attrep and Orth (1993) examined whether the Ir anomaly and the mass extinction at the D–C boundary was related to an extraterrestrial impact. They note that even though this Ir anomaly is found within a bed of black shale, known as the Hagenberg shale, globally, the stratigraphic position of this shale relative to the D–C boundary varies between sites. In France, the black shale is located about 20 cm under the D–C boundary whereas it lies immediately below the boundary in China. The D–C boundary Ir anomaly is comparable in magnitude to those of other major extinction boundaries; however, shocked quartz, microtek-tites and other indicators of extraterrestrial impact are generally absent. Since all the anomalies are found at redox and facies boundaries it is more likely that they resulted from a sudden change in the palaeoredox conditions and facies. This is supported by the fact that the black shales represent reducing and anoxic conditions, where there is an absence of most living organisms, and by high contents of organic carbon, chalcophile elements, sulphur and pyrite. Surrounding rocks associated with the shale are often well oxygenated and have normal marine fossils. Ir anomalies are most commonly found at the basal and/or upper boundary levels suggesting that redox changes at the boundaries played a more important role than did lithologies in controlling the distribution of Ir anomalies in these sections.

The Ir anomaly at the K–T boundary does contain many if not all of the other diagnostic signatures for an extraterrestrial impact. However, one of the main problems encountered when interpreting Ir and PGE abundances at the K–T boundary is the assessment of the extent to which elemental mobility obscures primary geochemical signatures. Although all PGE are mobile during weathering and diagenesis, some are more resistant to remobilisation from the original carrier phase. Evanc *et al.* (1993) found that the general order of PGE mobility during weathering is Pd > Pt > Rh > Ru > Os = Ir. Pt and Pd can form amine and purine organic complexes so it is possible that they can form complexes with free carboxylic acid and with soil organic matter. Since Ru and Ir are the least organically associated PGE, and also most resistant to chemical influences in nature, they are considered the most useful indication of the source for anomalies associated with extraterrestrial impacts.

Ni-rich spinel and spherule layers suggest that there is little doubt that the late Eocene was marked by a major extraterrestrial impact. Ir anomalies are also present within late Eocene marker beds. Montanari *et al.* (1993) examined Ir anomalies of two apparently complete and continuous late Eocene pelagic sequences. Both sequences are calibrated by both planktonic biostratigraphy and magnetostratigraphy. The core sample, from Italy, has also been calibrated by radioisotopic dates on biotite using K/Ar, Rb/Sr and Ar/Ar methods and U/Th/Pb dating on zircon and monazite from interbedded volcanic ash layers. Spherules were not found in these samples, even though Ir anomalies were present, and Montanari *et al.* (1993) suggest that the spherules were altered and obliterated by diagenetic processes. The lack of spherules and other diagnostic indicators show that the presence of Ir anomalies is probably insufficient evidence in itself to diagnose an extraterrestrial impact event. In the case of the late Eocene, however, the strong evidence elsewhere highlights that a bolide event probably did occur at this time and that Ir anomalies in horizons of this age are a product of this event. However, biological and chemical factors that can also produce Ir anomalies must always be eliminated before Ir alone can be taken as a reasonable indicator of an extraterrestrial source.

Zircon as an indicator for extraterrestrial impacts

Zircon is ideally suited as an indicator of shock-metamorphic impact events because of its transparency, lack of significant twinning or well-developed perfect cleavages and uniaxial optical character. Zircon is also more refractory than quartz and is therefore better able to resist subsequent thermal annealing of shock features in addition to being datable using the U–Pb isotopic method. Bohor *et al.* (1993) state that although only irregular radial cracking and open planar fractures had been reported in zircon from shock-impacted environments, these features were very similar to those developed tectonically in zircon from mylonitic terranes. With weathering, zircon decomposes into baddeleyite and silica from highly shocked impact glasses, and it exhibits a shock-induced displacive transformation to the scheelite structure before breaking down into mixed oxides. Bohor *et al.* (1993) found that impact-shocked zircon displays PDF when exposed to etching, and that shock-induced textures can be identified from distal ejecta, impact glasses and target rocks at known impact sites. The textures observed in zircon range from PDF to combined PDF/granular to fully developed granular (polycrystalline) texture and incipient melting. The described sequence is assumed to represent increasing shock pressure during impact (Bohor *et al.*, 1993).

Deformation effects have been identified in zircon crystals from impact breccias or shock-metamorphosed basement rocks at a number of confirmed impact structures and also at several K–T boundary sites. Bohor *et al.* (1993) found that etching of the zircon grains revealed a series of shock-induced features that suggest a textural response to progressive increases in impact shock pressure. The most common deformation features consist of planar features and strawberry or granular texture. The series of shock features extends from no shock features at all to those showing a continuous gradation from PDF alone, through PDF combined with granular texture, to a well developed granular texture and incipient melting phenomena.

Leroux *et al.* (1999) attempted to shock-deform zircon experimentally at shock pressures of 20, 40 and 60 GPa to establish the mineral behaviour under such conditions. They found the following behaviour.

(1) Deformation in the 20 GPa sample, and partially in the 40 and 60 GPa samples, includes both brittle and plastic deformation. Brittle

deformation mainly consists of thin, open cracks and are interpreted as micro-cleavages as the result of shock-induced shear stresses produced during the compression stage. Plastic deformation is indicated by abundant screw dislocations in a characteristic glide configuration. The large density of dislocations at crack tips suggests that plastic deformation is initiated by the micro-cracking process.

(2)   PDFs were found in the 40 and 60 GPa samples. They consist of thin planar defects filled with amorphous material of zircon composition. The formation of these PDFs is likely to occur at the shock front.

(3)   In the 40 GPa sample, a fraction of the zircon was converted to a high-pressure phase having a scheelite crystal structure. This phase deformation was complete in the 60 GPa sample. The phase transformation was found to be displacive. With regard to the shock process, it was established that PDF formation is an earlier process than transformation to a scheelite structure.

(4)   The optically resolved planar features in zircon are tentatively explained as micro-cleavage.

One of the main problems with using zircon as an indicator of extraterrestrial impacts is that it is much less abundant than other shockable minerals (Gucsik *et al.*, 2002). However, the reverse is true when attempting to examine impact features on extraterrestrial bodies such as meteorites and the Moon where quartz is in low abundance (Bohor *et al.*, 1993).

U–Pb isotopic dating can be applied successfully to zircon because it incorporates a small amount of uranium upon crystallisation, has one of the highest thermal blocking temperatures, is resistant to weathering and does not anneal under the same conditions as other minerals. U–Pb analysis of zircons and dating of the 'fireball' layer of the K–T boundary layer in the Western Interior of the USA was undertaken by Krogh *et al.* (1993). They found a limited range of primary ages from the target area rocks because of a lack of zircon in the primary rock. They determined an impact age of 65.5 ± 3 Ma, which concurs with previously established dates for the K–T boundary.

U–Pb dating of zircons has also provided strong evidence that the Chicxulub crater was the impact site for the K–T boundary event. Kamo and Krogh (1995) compared dated zircons from the impact site with zircon ejecta found at the K–T boundary in Colorado and Haiti. All the zircons have source ages of 545 ± 5 Ma, which is the age of some of the basement rock into which the crater was formed. The U–Pb dating technique is regarded as very reliable because it has two independent parent–daughter decay schemes permitting evaluation of both closed and open system behaviour (Deloule *et al.*, 2001). U–Pb ages, shock features, similar Pb loss patterns and the similar time of impact indicated by most of the zircon analysis conducted in the Kamo and Krogh (1995) study provide a temporal and genetic link between the K–T Ir-rich 'fireball' layer and the Chicxulub crater.

Isotopes as indicators of extraterrestrial impacts

While large bolides striking Earth often leave large impact structures along with substantial amounts of widely spread ejecta, small bolide impacts are more difficult to identify. In these situations identification of contaminates due to or from an extraterrestrial source is one of the most important means of identifying a bolide event. These contaminants or alteration signatures occur in impact glasses, melt rock and breccia. A strong enrichment in siderophile elements in impact melts is usually indicative of either chondritic or iron meteoritic components as these have a much higher abundance than terrestrial rocks. Achondritic projectiles are more difficult to detect as they have a significantly lower abundance of siderophile elements. Commonly used siderophile elements are Nickel (Ni), Cobalt (Co), Chromium (Cr) and their inter-element ratios. Cr enrichment and low Ni/Cr or Co/Cr ratios can be used to distinguish between iron and chondritic projectiles as chondrites usually have higher Cr abundances than iron meteorites. One major problem with this identification technique, however, is the fact that Co, Cr and Ni are common minerals and that enrichment may be natural (Koeberl and Shirey, 1997).

As noted, the PGE are useful in identifying meteoritic components. These elements occur at concentrations several orders of magnitude higher in meteoritic rocks compared to terrestrial rocks. Ideally though, indigenous concentrations of these elements should be subtracted from the concentrations identified in the melt rocks to yield meteoritic abundance ratios that are specific to the impact event. There are several reasons why this is difficult practically; these include the target rock not always being exactly known or indigenous PGE concentrations being variable or very low. Fractionation and hydrothermal processes are also problematic when establishing meteoritic origin (Koeberl and Shirey, 1997).

Koeberl and Shirey (1997) suggest that isotopic compositions of Os, together with Rhenium (Re) and Os abundance, are useful in determining meteoritic origins. For other radioactive systems, parent and daughter elements are incompatible in mantle rocks. As such, the Re–Os system is unique. Os is strongly retained in the mantle during partial melting of mantle rocks and remains in the residue, while Re is moderately incompatible and is, therefore, enriched in the melt. This results in high Re and low Os concentrations in most crustal rocks. Both the present day mantle and meteorites have low ^{187}Os/^{188}Os ratios.

Although Re and Os content is much higher in chondrites and iron meteorites than the mantle, the Re/Os ratio is indistinguishable from that of meteorites. In crustal rocks, the abundance of Os is much lower than in the mantle and the Re/Os ratio is significantly different between mantle rocks and meteorites. Thus, differences in the Re/Os ratio of meteoritic and crustal rocks lead to differences in the increase of the radiogenic isotope ^{187}Os. The ^{187}Os/^{188}Os ratio of meteorite and mantle rocks changes relatively slowly with time, whereas the crustal ^{187}Os/^{188}Os ratio changes relatively rapidly with time due to the high Re concentrations. These features result in an identification tool for meteoritic rocks. Sudden decreases in ^{187}Os/^{188}Os ratios have been found at the K–T boundary and other deposits following extraterrestrial impacts. It seems reasonable that this method can be used to provide a better understanding of the mixing processes that occur between target rocks and impactors. Used with care, the Re–Os dating method can provide a diagnostic tool for establishing an extraterrestrial origin similar to that of shock metamorphism (Koeberl and Shirey, 1997).

MacDougal (1988) combined precise documentation of the strontium isotopic composition of ocean water through time and an understanding of the atmospheric geochemical effects that accompany large extraterrestrial impacts to analyse events at the K–T boundary. MacDougall suggested three possibilities might explain the anomalously high strontium concentrations at the K–T boundary. First, it is possible that the elevated ^{87}Sr/^{86}Sr ratio was a result of dissolution of the bolide in seawater. Second, it is possible that the impact ejecta (both vapour and solid) were a major source of strontium, although it is not likely as meteorites would not have a sufficiently high strontium composition to produce the recorded anomaly. Third, the atmospheric chemical effects of the impact from a large projectile could result in the production of large amounts of nitrogen oxides due to shock heating. This would result in extremely acidic precipitation near the impact centre and also globally over a more extended time span. Acid rain is likely to increase continental weathering with the result that continental strontium to the oceans is increased. MacDougal (1988) suggested that the latter mechanism is the most probable cause of the enhanced ^{87}Sr/^{86}Sr ratio discovered at the K–T boundary. Volcanism can also potentially produce acid rain and result in the same processes. However, this is likely to be a long-term effect, rather than a distinguishable peak as is the case at the K–T boundary.

Isotopic chemostratigraphy and chronostratigraphy on sites in Antarctica and Australia were used by Retallack et al. (1998) to determine whether the Permian–Triassic (P–T) boundary, which is associated with one of Earth's greatest oceanic mass extinction events, was associated with an extraterrestrial impact. Microspherules, spinels and iridium anomalies have been reported from this time period but the validity of these findings has often been disputed. Possible impact craters are also either incompletely documented or do not correlate in age. Explanations for the event range from oceanic anoxia, oceanic CO_2, to overturn and eruption of flood basalts from the Siberian Traps. Retallack *et al.* (1998) found similarities in the isotopic composition of the boundary beds (6–15 cm thick claystone breccia) at locations in both Antarctica and Australia. These beds are interpreted as redeposited soils as they do not contain spherules or other indicators of impact ejecta. However, the claystone breccias do contain rare grains of shocked quartz and a small iridium anomaly was identified. Unfortunately, the Retallack *et al.* (1998) study did not shed any further light on the mass extinction event at the P–T boundary but a 'hint' of an impact was revealed.

Conclusion

Asteroid and significant meteorite impacts with Earth are uncommon events and rarely taken seriously in terms of any risk assessments. The prehistoric record shows that truly catastrophic events have occurred in the past and no doubt will do so in the future. There is little doubt that this hazard needs to be taken more seriously but until such an event occurs, even if it only causes relatively minor damage or a few deaths, governments and public authorities are unlikely to ever give this threat serious consideration. The enormity of the potential consequences and their very low frequency of occurrence are no doubt instrumental in the propagation of this attitude. However, the unquestionable origin of much of the evidence such as shock structures in combination with element anomalies and physically imposing features, such as craters, show us that mitigation against this hazard will require very sophisticated technology and be of great expense.

10

Extreme events over time

The physical processes operating during extreme events (or natural hazards when they affect people) are now relatively well understood. There are also a broad array of approaches and methodologies available for unravelling the long-term history of extreme events from natural records. As a corollary, sufficient data has now been collected on the long-term behaviour of these phenomena to allow more realistic assessments of their frequency and magnitude compared to that achievable from short historical records alone. To date, however, few risk assessments, especially those dealing with atmospheric hazards utilise this long-term data.

One of the most interesting aspects of the long-term records is that they often show that short historical records do not accurately reflect the behaviour of a hazard over longer periods of time. Hazard behaviour appears to differ between the two temporal records in terms of variability and stationarity or non-stationarity. Short historical records often do not display the full variability of a hazard and they also often give the impression that the hazard displays stationarity. Of course a short historical record may reflect the true behaviour of a hazard, but a risk assessor can never be sure of this unless comparisons are made with actual records of the longer term behaviour. Many of the long-term records discussed in earlier chapters show, however, that short historical records poorly reflect the true nature of the hazard. As a consequence, problems can arise when risk assessors extrapolate from historical records and assume that the return interval of extreme events is much longer than is realistically the case. In such situations people and property are unnecessarily exposed to hazards resulting in increased community vulnerability and risk.

Short historical records can also give the impression that natural hazards occur in a stochastic fashion over time. However, long-term records regularly show that natural events, including natural hazards, display serial correlation or cyclicity over time. Serial correlation is a condition where the occurrence of an event in a time series is dependent upon or related to the previous event and it suggests cyclicity in the record. The apparent 'white noise' or randomness of hazards evident in short historical records is in reality part of a longer term trend towards cyclicity. If hazards display serial correlation over time, then statistical techniques based upon the random occurrence of events will not accurately estimate the probability of hazard occurrence. In other words, if a hazard does follow a pattern of serial correlation then the probability of its occurrence must change with time. Most assessors of risk assume that the probability remains constant over time so the probability of an event occurring during the next 50 or 100 years is seen to be exactly the same as during any 50–100 year period. This assumption derives from the hypothesis that the behaviour of a hazard over the short historical record is a true reflection of its longer term behaviour. However, if serial correlation is recognised in a longer term record then the probability of an event occurring over the past 50 years is likely to be different from its probability of occurrence over the next 50 years, especially where the cycle of hazard occurrence is greater than 100 years.

Serial correlation is evident in many of the long-term records discussed in earlier chapters. This chapter briefly reviews the characteristics of these long-term records and then presents the statistical evidence to demonstrate that these time series display serial correlation and hence cyclicity over time.

Atmospherically generated extreme events

Despite the assumptions made by many who undertake risk assessments of the tropical cyclone hazard, there is little evidence, if any, to support the notion that these events occur randomly over the long term. Indeed, several studies have found that there have been discrete periods of time when these events tend to cluster. Liu and Fearn's (2000, 2002) analysis of washover deposits along the Louisiana and Alabama coasts show that the frequency of severe hurricanes has not been constant over time. A 2400 year period of increased cyclogenesis occurred here between 3200 and 1000 ^{14}C years BP. Eleven of the twelve high-intensity hurricanes that struck this region over the last 4800 years occurred during this period and only 1 occurred over the past 1000 years. Furthermore, no hurricanes of the intensity registered in the prehistoric record have struck the region during the past 130 years. The 1600 year period prior to 3200 years BP was also relatively quiescent. A similar story is evident for the long-term Chinese record of tropical cyclones. This record is an historical one, but one that is obviously much longer than occurs in most other countries. Its length, however, highlights that here too there have been periods when the frequency of high-intensity events has increased. Liu *et al.* (2001) showed that clusters of increased cyclone frequency occurred between AD 1660–1680 and 1850–1880. At other times there were quiescent periods in the record. With a record of this length it is possible to obtain a reasonable reflection of the nature of this hazard for it displays non-stationarity over the longer term. These periods of regime shifts can then be incorporated into any risk assessment of the hazard and a more realistic view of the level of exposure of people and infrastructure along with community vulnerability can be made.

Hayne and Chappell (2001) have suggested that the occurrence of tropical cyclones along parts of the Great Barrier Reef have not shown any appreciable change in frequency over the past 5000 years. Nott and Hayne (2001) subsequently showed that these coral shingle ridge deposits were most likely emplaced by the most extreme cyclones to strike the region and hence preferentially record only this spectrum of the cyclone climatology. Therefore, on average, it would appear that the most extreme tropical cyclones in this region have not shown any appreciable change in frequency of occurrence and, therefore, could be regarded as belonging to the one hazard regime over this extended period of time. However, while these events have occurred on average every 200–300 years, there is considerable spread in the data showing that the time interval between actual events has varied from 50 to >500 years. More importantly though, and despite the fact that these events at present appear to display randomness, their frequency of occurrence is an order of magnitude higher than that suggested by the historical record. Hence, the historical record is a poor reflection of the true behaviour of the most extreme tropical cyclones in this region. Again, the prehistoric record shows that the historical record does not display a complete pattern of the true variability of this hazard. Any risk assessments, therefore, that rely solely upon the short historical record here are almost certainly likely to underestimate the risk of the hazard.

Of course any interpretations of the nature of the hazard are dependent upon the scale and resolution of the record. The fact that no extreme cyclone events have occurred at many of the sites analysed by Nott and Hayne (2001) over historical times (i.e. the last 130 years) suggests that if the historical record could be extended to about 300 years then it would be likely to show at least one and maybe two extreme cyclone events during this period. It may also be the case that these two events occurred over a relatively short time interval. Nott (2003) showed that natural cyclone records spanning the last 200 years near Cairns, North Queensland reveal that two extreme tropical cyclones occurred between about AD 1800 and 1870, and none have occurred since (being the period covered by the historical record). The radiocarbon chronology is not sufficiently precise to determine the exact years in which these events occurred; however, the fact that they definitely occurred within a seventy year period, and probably within a shorter time interval, suggests that this may have been a period of enhanced cyclogenesis which has not occurred since. The implication is that if long-term tropical cyclone records of higher resolution could be obtained for this region, particularly in terms of covering the entire range of cyclone magnitudes, then randomness and stationarity may not be apparent. Indeed, the fact that these extreme cyclones did occur just prior to the start of the historical record provides a test of the historical record and shows that the assumptions about the frequency distribution of tropical cyclones in this region are incorrect.

Long-term records of terrestrial floods certainly show that these events cluster into phases of heightened and lesser activity. The luminescence line record within corals of the Great Barrier Reef shows that terrestrial floods were considerably larger between AD 1640 and 1760 and AD 1870 and 1910 with periods of relative quiescence between AD 1830 and 1870 and again for much of the 1900s. The gauged flood record for this region starts around AD 1916, and, therefore, does not capture the true variability of flooding. At longer time scales, the plunge pool record from the Northern Territory, Australia also shows variability with heightened flood activity between 8000 and 4000 years BP and lower flood activity (indeed flood magnitudes are about 20% lower on average) over the past 4000 years. While this is not critical for most planning timescales, it is critical in this region because the catchment from which some of these records are derived is mined for uranium, and tailings from this process are expected to be stored in a stable environment for at least 1000 years. Nott and Price (1999) showed that the enhanced phase of flooding between 8000 and 4000 years BP corresponded to a different climatic regime, and one which is thought to be similar to that expected under an enhanced greenhouse climate. It is now clear that small shifts in average climatic conditions can result in substantial shifts in flood magnitude and a return to a phase of greater flooding is entirely possible within the next few decades or century in the Northern Territory. So, even though the timescale of prehistoric flood regime shifts may be relatively large, they are certainly within the time frame, given the prospects of future climate change, to be taken seriously when considering the safe storage of uranium tailings.

Ely *et al.* (1993) and Knox (1993, 2003) have likewise demonstrated a clear relationship between shifts in climate state and the magnitude of floods. Ely *et al.* (1993) showed that floods in the southwest USA were relatively frequent between about 5800 and 4200 years BP and after 2400 years BP, except for the period between 800 and 600 years BP. An absence of, or rarely occurring, floods dominated the period between 4200 and 2400 years BP. Knox (2003) concluded that floods in the Upper Mississippi valley were large between 600 and 300, 1000 and 750, 1800 and 1500, and 2500 and 2200 years BP. Either moderate or small floods dominated the episodes between these periods. The key point about these studies of long-term flooding history in the USA is that flood magnitudes and frequencies have been highly variable over the past 7000 years. The causes behind these variations in flood frequencies have been shifts in the climatic state where large floods have tended to occur during warmer climatic intervals associated with shifts in the position of the jet stream axis over the USA (Knox, 2003).

Of all the atmospherically-generated extreme events, long-term drought records probably provide the best insight into the true behaviour of an extreme event or hazard over time. Drought records generally have a higher resolution than that able to be obtained for many other hazards or extreme events because they can be recorded in annual sediment layers, tree rings and speleothem layers. The higher resolution of drought records permits a more definitive test of the hypothesis of randomness of a natural extreme event or hazard over time. Indeed, the long-term drought records unquestionably show that these events display variable frequency over time.

Using $\Delta^{13}C$ records from peat cellulose, Hong *et al.* (2001) showed that eight, multi-decade to multi-century, drought periods occurred between 2200 BC and AD 1200 in northeast China. After this time, the climate returned to wet and cold conditions between AD 1200 and 1800, a period during which the frequency or severity of droughts was much lower. Laird *et al.* (1996) used diatoms to reconstruct a remarkable subdecadal record of lake level variations in the USA that clearly depicts a dramatic change in drought regime after AD 1200. Prior to this time droughts dominated the region following which few severe droughts are apparent in the record.

Verschuren *et al.* (2000), in their reconstruction of the drought and rainfall history of tropical east Africa over the past 1100 years, showed that three distinct drought episodes occurred in the early 1300s, between approximately AD 1450–1550 and around AD 1700. Prior to AD 1200, drought appears to have been consistently more severe. In east India, Nigram *et al.* (1995) found a 77 year cycle of severe droughts over the past four to five centuries based upon foraminifera records. Similarly, Mullins (1998) found five distinct phases of cooler and drier conditions in Cayuga Lake, New York over the past 10 000 years that were interpreted as episodes of severe drought that lasted several centuries. Mullins (1998) suggests that the occurrence of these drier episodes followed a cycle with a period of 1800 to 2200 years. Liu *et al.*'s (1997) stalagmite record from Shihua Cave, Beijing showed seven major episodes of drought over the past 1100 years and the occurrence of several climatic cycles with periodicities of 136, 50, 16–18, 11 and 5.8 years. There was also evidence for possible millennial scale variability.

Dean (1997) likewise recognised severe droughts with a 400 year periodicity in the 10 000 year long varve record of Elk Lake, USA. This periodicity is centred at 1200–1000, 800–600 and 400–200 years BP and these episodes are interpreted to represent episodes of windier and dustier conditions compared to the period between 3000 and 1500 BP. Two other cycles of aeolian, and hence drought activity, were recognised with periods of 1600 and 84 years. Based upon this record, Dean (1997) stressed that the 20th Century droughts of the USA are not representative of the full range of droughts that have occurred over the past 2000 years. Here, multi-decadal droughts of the late 13th and 16th Centuries were more prolonged and severe than those of the 20th Century and few, if any, major droughts have occurred in the periods between these episodes of severe drought. The long-term record here also suggests that there was a regime shift in droughts after the 13th Century. Droughts prior to the 20th Century were at least decades in duration, whereas after this time droughts were a decade or less in duration. As Woodhouse and Overpeck (1998) emphasise, the short historical records of droughts from the USA, Australia and Africa are often not a reliable guide to drought climatology and 20th Century records are less than ideal as a guide to future events.

Non-atmospheric events

The resolution of the natural long-term records of earthquakes, landslides, volcanic eruptions and asteroid impacts with Earth is lower than the long-term records of atmospheric events such as droughts and floods. These events also tend to occur less frequently and there is usually a substantial period of inactivity between them. Hence, they do not lend themselves to being registered in annual records as readily as some of the atmospherically generated events. This is because the considerable time between events, such as four to five centuries, is often longer than the lifespan of individual trees and sometimes even lakes, and the accumulation period of some speleothems. This does not mean that these events cannot be recorded by these systems. As shown in Chapters 6–9, non-atmospheric events have been recorded in the form of reaction wood in trees from landslide events, and drowned forests and sediment accumulations in lakes from subsidence and mass movements. The real value of the prehistoric records of the non-atmospherical events is that they provide a much more realistic guide of the return interval of these events rather than a test of their randomness of occurrence over time. The prehistoric records often do not provide a test for the latter because the chances are reasonably high that these events will not have occurred, at least at extreme magnitudes, within the timeframe of a short historical record. There is also a certain level of conditional probability and acceptance of this fact by risk assessors that the longer it has been since the last event the sooner it is until the next one. This is because time is required for pressure and stress to build for volcanic eruptions and earthquakes, respectively, after the last event. The same is true to a certain extent with landslides, where these events may be caused by the accumulation of weathered bedrock or saprolite on a slope. Once this material fails and moves downslope, it takes some time before the fresh bedrock is able to weather to a sufficient degree so that a substantial saprolitic mantle can accumulate. This then is really a function of the sensitivity of the landscape, particularly in non-seismic tropical regions where many landslides can be generated by intense rainfall events saturating a thick mantle of weathered material on a steep slope (Thomas, 2004).

The palaeoseismic record allows us to gain an appreciation of the magnitude and frequency of events likely in a region, especially when such events have not occurred for some considerable time. Meghraoui *et al.* (2003), for example, recognised that the absence of major earthquakes along sections of the Dead Sea Fault in Syria over the past 830 years is not a reasonable guide to the likely future occurrence of earthquakes. They found that three major events of magnitude M_W 7.0 and 7.5 occurred over the last 2000 years. The average recurrence interval of 550 years for large earthquakes at this location shows that the fault zone is not inactive and also that the next large earthquake may be imminent, or possibly overdue. Such detailed conclusions would not have been possible from the historical record.

Palaeoseismic records from the San Andreas fault, California do suggest that recurrence intervals for large earthquakes here may not have been consistent over time. Fumal *et al.* (2002a) and Biasi *et al.* (2002) interpret these records to show that there may have been a relative acceleration of earthquake activity between the early AD 600s and the early 800s followed by a lower rate around AD 1500 and then by a higher rate after this. Biasi *et al.* (2002) suggest that strain transfer due to earthquakes along other sections of the same fault or other faults may influence the timing of earthquakes. Recurrence intervals can be affected by the accumulation and release of fault normal stresses at restraining bends in a fault causing the recurrence intervals of earthquakes along nearby faults to be different. The stress release and accumulation may be influenced by the occurrence of earthquakes on these nearby faults. The two systems therefore interact. Strain conditioning of the recurrence interval may also be another possibility. Biasi *et al.* (2002) suggest that earthquakes might occur at regular intervals for some period of time but stress may not be completely released and it gradually accumulates. They also suggest that the physical properties of the fault may vary with time such as the characteristic size and spacing of asperities which may influence the ease with which ruptures initiate and also the nature of fault strength. Whatever the cause, palaeoseismic evidence in the USA and elsewhere such as the eastern Mediterranean (Pirazolli *et al.*, 1996; Biasi *et al.*, 2002) suggests that earthquakes may not occur entirely randomly over time and that events may increase in frequency at times and decrease during other periods.

Theoretically, if earthquakes can cluster over time then so too could seismically generated landslides. Crozier *et al.* (1995) noted several episodes of regionally synchronous landslide events in the Taranaki region of New Zealand's North Island between 1200 and 1400 years BP and around 30 000 years BP. These events were triggered by seismic events as opposed to episodes of heavy rainfall. Crozier *et al.* did not discuss whether these events represent clusterings of landslides over time but the interesting conclusion from their study was that none of these multiple deep-seated landslides had occurred during the 150 years of European settlement. Clearly, without the prehistoric record, the obvious conclusion to draw is that this region rarely, if ever, experiences large landslides. However, the long-term record demonstrates otherwise.

Tsunamis may also cluster, but there is little definitive independent evidence to support this. However, like the Taranaki landslide events, tsunamis in some locations can often be absent within short historical time frames. Australia is a case in point as this country is rarely impacted by sizeable tsunamis and certainly not within the time frame of European occupation (last 230 years). Nott and Bryant's (2003) study of the palaeotsunami record in Western Australia, being Australia's most tsunami-prone coast, showed that sizeable tsunamis had overtopped cliffs 25–30 m above sea level only several hundred years ago. The ages of shell deposits on headlands and within mixed sand and gravel deposits along the coast of Western Australia suggests that the region has experienced several large tsunami events over the past 5000 years. Most of these appear to have occurred over the last 2000 years (Fig. 10.1) with a substantial gap between events in the period around 3000–5000 years BP. Part of this chronology may be biased by the preservation potential of shells in deposits along this coast, such that more recent deposits are more likely to be preserved. However, the palaeorecords highlight that tsunamis have occurred in the past and that they were substantially larger than the small events that have occurred historically.

Quantitative evidence for non-randomness

A number of different statistical techniques can be used to test whether a time series displays randomness or is serially correlated. Serial correlation in a time series infers that the series displays cyclicity. The random occurrence

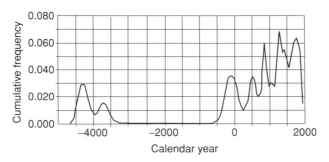

Figure 10.1. Cumulative probability frequency of ^{14}C (uncorrected) ages for the northwest coast of Western Australia (from Nott and Bryant, 2003).

of an event, as mentioned in Chapter 1, suggests that events in a time series occur independently of each other and, therefore, are not cyclic. Figures 10.2 and 10.3 show the results of two separate statistical analyses for serial correlation of several of the long-term high-resolution records of the natural hazards discussed in earlier chapters. Each of the records is an annual register of century to millennial length. Some natural hazards, such as tropical cyclones, do not occur annually and cannot, therefore, be analysed in this manner. However, some of the forcing mechanisms responsible for the frequency of occurrence of these hazards (Nichols, 1984) do occur annually and can be included in the statistical analysis. These records include sea-surface temperatures (SST), North Atlantic Oscillation (NAO), Pacific Decadal Oscillation (PDO), El Niño Southern Oscillation (ENSO) and rainfall.

Figure 10.2 displays the results of a Runs test on a variety of natural hazards such as North American droughts as recorded by tree rings, stream discharges, rainfall years from speleothem layers in China, Scotland and Australia, coral layer density, along with SST, NAO, PDO and ENSO. The Runs test examines the number and length of increasing and decreasing runs or trends in a time series. For example a time series may be composed of a sequence where the trend is increasing (e.g. 1, 2, 3, 4, 5) and then a shorter run of decreasing numbers (e.g. 7, 6, 5). The number and length of these individual runs is compared to that expected from a series of random events. The null hypothesis is that the time series is random and the test statistic is represented by the Z score. Where the Z score exceeds approximately 1.9, the null hypothesis can be rejected at the 0.05 significance level. Three separate tests were also undertaken on a series of randomly generated numbers to highlight the pattern expected if the time series of natural hazards and events were truly random. Each of the time series was also analysed at cumulative incremental time lengths with 50–100 year intervals. In other words, the Runs test was undertaken on the first 50 or 100 years of record,

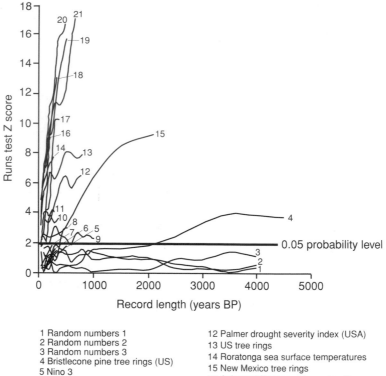

Figure 10.2. Results from Runs test for randomness of various long-term, high-resolution time series. (Data from 4, Hughes and Graumlich, 1996; 5, Cook, 2000; 6, J. Nott; 7, Cleaveland, 2000; 8, Cook, *et al.*, 2002; 9, Meko *et al.*, 2001; 10, Isdale *et al.*, 1998; 11, Graumlich *et al.*, 2003; 12, Cook, 2000; 13, Cook *et al.*, 1996; 14, Linsley *et al.*, 2000; 15, Grissino-Mayer, 1996; 16, 19, 20, Lough and Barnes, 1997, 2000; Chalker and Barnes 1990; 17, Tan *et al.*, 2003; 18, Biondi *et al.*, 2001; 21, Winter *et al.*, 2000.)

then the first 200 years and first 300 years and so on. In the vast majority of cases the Z scores exceeded the 0.05 probability level suggesting that these long-term records are serially correlated (Fig. 10.2). The Z scores also increase with time or length of record. Some of the shorter length records (Chillagoe speleothem layers, Sacramento River discharge and White River discharge) did not reach the critical Z score of 1.9 (or 0.05 probability); however, the steep upward trend of the curve shows that if each of these records were 50–100 years longer they would invariably display serial correlation. The Z scores of the random

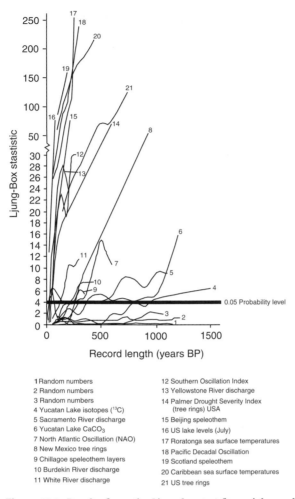

Figure 10.3. Results from the Ljung-box test for serial correlation. (Data from 4, 6, Hodell *et al.*, 1995; 5, Meko *et al.*, 2001; 7, Cook *et al.*, 2002; 8, Grissino-Mayer, 1996; 10, Isdale *et al.*, 1998; 11, Cleaveland, 2000; 12, Allan *et al.*, 1996; Stahle *et al.*, 1998; 13, Graumlich *et al.*, 2003; 14, Cook, 2000; 15, Tan *et al.*, 2003; 16, Cook *et al.*, 1996; 17, Linsley *et al.*, 2000; 18, Biondi *et al.*, 2001; 19, Proctor *et al.*, 2000; 20, Winter *et al.*, 2000; 21, Cook *et al.*, 1996.)

numbers did not exceed 1.9 despite the substantial length of record (equal to 4000 runs or years if one run equals a year). The curves of the random numbers also do not show any substantial increase with increasing length of record suggesting that they would probably never exceed the critical Z score of 1.9.

The Ljung-box test, another statistical test of randomness, was also undertaken on many of the same long-term hazard and natural event time series. The Ljung-box test calculates a correlation coefficient for different lags in the time

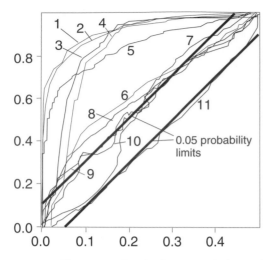

1 Sanctuary Is. annual coral band density

2. Agincourt reef annual coral band density

3 Pacific Decadal Oscillation

4 White River discharge

5 Roratonga sea surface temperatures

6 Palmer Drought Severity Index (USA)

7 Southern Oscillation Index

8 New Mexico tree rings (precipitation)

9 Burdekin River discharge

10 Chillagoe speleothem layer thickness

11 North Atlantic Oscillation

Figure 10.4. Results from cumulative periodograms for serial correlation. (Data from 1, 2, Lough and Barnes, 2000; Chalker and Barnes, 1990; 3, Biondi *et al.*, 2001; 4, Cleaveland, 2000; 5, Linsley *et al.*, 2000; 6, Cook, 2000; 7, Allen *et al.*, 1996; Stahle *et al.*, 1998; 8, Grissino-Mayer, 1996; 9, Isedale *et al.*, 1998; 11, Cook, *et al.*, 2002.)

series. A lag is a comparison or relationship between events of a certain distance apart in the sequence. For example a lag of two compares events separated by one event (or every second event), a lag of three compares events separated by two events and a lag of four or five are comparisons between events three and four events apart respectively, throughout the length of the time series. The time series examined using the Ljung-box test were also analysed over various length–time intervals and as seen in Figure 10.3 all of the curves except for the random numbers also show steep upward trends and exceed the 0.05 probability level. Like the Runs test, the Ljung-box test has a null hypothesis that the time series is random. When the test statistic exceeds the 0.05 probability level there is a 95% chance that the time series is not random and therefore is serially correlated or shows auto correlation at certain lags. Again the null hypothesis of randomness in the time series examined can be rejected after 100–200 years of record; indeed in some cases it can be rejected after just 50 years of record. The time series were also examined using cumulative periodograms (Fig. 10.4), which test for the probability that the events in a time series represent 'white noise' or randomness. The time series can be regarded as non-random when the line extends outside of the 95% confidence limits. As with the Runs and Ljung-box tests, the cumulative periodograms highlight that time series of both natural hazards and natural events such as SST, PDO, NAO, ENSO and rainfall regularly display serial correlation over time. The first part of the record (∼100 years), however, commonly appears as white noise suggesting that this

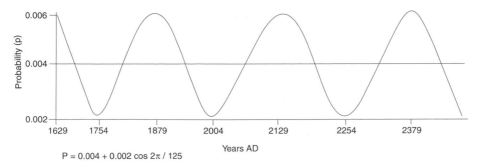

$$P = 0.004 + 0.002 \cos 2\pi / 125$$

Figure 10.5. Temporally variable probabilities of extreme magnitude tropical cyclones along the Great Barrier Reef. This schematic is an approximation to show that probabilities change over time. However, they may not change with the precision suggested by this cosine curve and another curve may be more appropriate. Uncertainty margins are not shown. Mean probability = 0.004 and probabilities vary between 0.006 and 0.002. The variation in probability will vary depending upon the amplitude of the curve chosen. Ideally the amplitude would be determined by fitting a regression curve to the time series data. This curve suggests a new approach or direction towards determining probabilities rather than a definitive method.

apparent randomness is really a subset of a longer term cycle or series of cycles.

The brevity of many historical records leads most risk assessors (particularly of atmospheric hazards) to believe that the apparent white noise of the short record is also a realistic characteristic of the hazard over longer time intervals. Unusually large magnitude events within short historical records are therefore assumed to be outliers. However, if the time series is not random, a more realistic interpretation is that the probability of a given magnitude event may change over time. At times, following a cyclic or quasi-cyclic pattern, the hazard probability will be higher than the mean probability and at other times the probability will be lower than the mean (Fig. 10.5). Nott and Hayne (2001) determined a mean return interval of 250 years for the most extreme tropical cyclones striking the east coast of Queensland, Australia based upon a 3000 to 5000 year record from seven separate sites. It is entirely possible that these extreme cyclones were occurring during periods when conditions were more conducive to their occurrence, because these conditions tend to occur cyclically rather than randomly. If this were the case then the probability of the most extreme events occurring will vary from 0.006 to 0.002 with a mean of 0.004 (Fig. 10.5). The periods when the most extreme events were occurring had at that time a probability of occurrence of 0.006. In between these periods the probability would have been lower.

While the records of tropical cyclones (both historic and prehistoric) in this region or anywhere are not annual and therefore cannot be analysed statistically for randomness, many of the mechanisms that influence tropical cyclone occurrence and abundance do display serial correlation over the long term. These mechanisms include the SOI, PDO and SST (Figs. 10.2–10.4). If these boundary mechanisms are not random then it is also unlikely that tropical cyclones will occur in a truly random fashion over time periods longer than the short historical record. Given that the preservation potential of the tropical cyclone palaeo-record in this region results in only the most extreme events being registered over time, it is logical that these events most likely occurred during phases of increased occurrence probability represented by the crests of the probability curve (Fig. 10.5). At these times these extreme events would have been equivalent to the 1 in 166 year event or $p = 0.006$ (annual exceedence probability, AEP). If an event of this magnitude were to occur during a period represented by a trough in the probability curve then it could be regarded as a 1 in 500 year event or $p = 0.002$. In this sense, these latter events are not outliers but events whose probability varies with time.

This same approach can be adopted for other kinds of hazards such as river floods. Phases of enhanced flooding in the Burdekin River, North Queensland, based upon the long-term coral luminescence record (see Chapter 3, Fig. 3.11), have a period of approximately 150 years. In this case large-magnitude floods occurring during the enhanced phase of flooding, or cycle peak, would have a probability of occurrence (AEP) of 0.0225 and during troughs an AEP of 0.0075. Statistical analysis of this flood record using the KPSS test for stationarity (Kwiatkowski et al., 1992) shows that it displays non-stationarity. Two approaches could be taken in determining the AEP of given magnitude floods in this situation. The first is as mentioned above where a mean AEP (0.015) of large-magnitude floods could be assigned to cover the entire record. Or alternatively, separate probability cycles could be developed for the enhanced phases and lesser magnitude phases each with their own mean AEP. The Burdekin River coral luminescence flood record is only 350 years in length and ideally a longer record might be better suited to the latter approach for estimating probability.

These approaches to assessing temporally variable probabilities for hazard risk assessment can only be regarded as preliminary at this stage. To be more realistic they will require uncertainty margins and a reasonable idea of the position of the present day in the probability cycle. The causes of the cycles are also unlikely to be as straightforward as those producing clusters of earthquake events. Cycles in the probability of occurrence of atmospherically generated hazards are likely to be a function of the interaction of many cycles, both within one variable and between different variables. At times the interaction of these cycles will tend to enhance the probability due to some amplification of the necessary conditions required for development of the hazard in question. For example, in the case of tropical cyclones along the Great Barrier Reef it is possible that peaks in the cycles of a number of variables (SST, PDO, SOI) coincide at an average frequency of 250 years (phase peaks rather than variables with cycles of identical periods). It is also likely that the cycle variability is considerable so no one cycle may persist over long periods of time. Rather, the time periods between phases of increased probability of hazard occurrence may vary. In this situation a mean cycle period is likely to be the most appropriate measure of the periodicity of the temporally variable probability with an uncertainty margin that reflects the natural variability displayed in the long-term record. Despite the potential difficulties in employing such an approach, it nonetheless allows for the recognition and adoption of the fact that the hazard probability can change with time and therefore more accurately reflects the return period of so-called outlier events. Identifying the last major phase of enhanced hazard activity from the palaeorecord will provide a reasonable estimate of the position of the present day in the cycle. The likelihood of occurrence of a hazard of given magnitude over the next 50–100 years can then be assessed. At times this may suggest the probability of occurrence over a planning period to be lower than otherwise expected and at other times this probability will clearly be higher. Adopting such an approach acknowledges that natural hazards and their causal mechanisms do not occur entirely randomly over time and there is to some degree an inherent cyclicity or quasi-cyclicity in their occurrence. This then logically suggests that their probability of occurrence must change over time also and therefore risks must likewise be temporally variable.

Incorporating palaeorecords into hazard risk assessments

When developing policies, planning authorities usually focus on those documents that summarise the state of the risk (all three components – physical characteristics and community exposure and vulnerability) rather than refer to the more technical scientific papers that deal with the physical characteristics of the hazard alone. Hence, it is incumbent upon those assessing community risks to a natural hazard to also address the accuracy of estimates of the frequency characteristics of the hazard. At the very least the risk assessors need to draw attention to the level of uncertainty associated with such estimates. By doing so, and as a corollary, any study addressing community risk could incorporate variable temporal probabilities and their uncertainty margins into measures of community exposure and vulnerability.

Planning often requires balancing economic gains and losses against the risk of the hazard occurring. The life expectancy of many modern urban developments is between 50 and 100 years. So estimates based upon a short historical

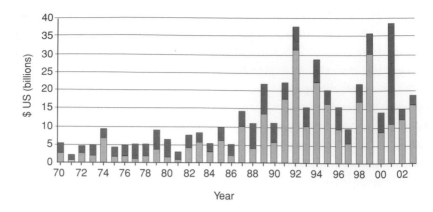

Figure 10.6. Insured losses from both natural and human-induced hazards (from Swiss ReInsurance).

record may predict an appropriate and safe position for a development with respect to hazard exposure if the hazard regime remains the same for the next 50 or more years. However, this cannot be predicted with any certainty until the nature of any hazard regime shifts and variable probabilities are known. This requires knowledge of the long-term variability of the hazard in question. Without such knowledge it is prudent to assume that a shift to a more active hazard regime is possible within the lifespan of the planning period. When, however, knowledge of the long-term variability is obtained and the variations in probabilities are determined, it is up to planners and policy makers to use their own judgments on where they intend to set the level of risk for community safety. This of course is always subject to a range of different economic and political forces. However, it will always be prudent to adopt the precautionary principle especially when it may be uncertain exactly where the present day lies on the probability cycle.

Future climate change and natural hazards

Globally, insured losses from natural hazards have increased substantially over the past 40 years (Fig. 10.6). While it is obvious that increasing population and community vulnerability will account for most if not all of this trend it will also be tempting for many to suggest that global climate change may be at least playing a role resulting in an increase in actual hazard events. Predictions from global climate models suggest that the magnitude and in some instances the frequency of atmospherically generated hazard events will increase. The critical question is knowing when this has occurred or is occurring and know-ing how to determine this. Assumptions of stationarity and randomness will lead many to think it is reasonable to look for changing trends in short historical records and if an increase is observed over the past few decades then this is likely evidence for the impact of climate change. However, such an approach could be very misleading because the natural variability of the particular hazard has not been decoupled from the observed trend. In other words it is important to ask whether the trend is simply part of a natural cycle in hazard behaviour. The critical test for increasing air temperature as a function of climate change has been to compare trends occurring over the last century to that recorded, both from proxy and other records, of temperature changes over multi-century to millennial scales. In particular, much focus has been upon the extent of temper-ature changes during the medieval warm period compared to the last century. Hence comparisons with the long-term record have formed the critical test for assumptions about recent global changes in air temperature.

The same is true when attempting to determine whether human-induced cli-mate change is playing, or has played, a role in changing the frequency and magnitude of natural hazards. It is difficult to decouple human-induced change from the natural variability of the hazard without knowledge of the long-term record. Therefore, comparisons with the long-term record are necessary to deter-mine *if* climate change is playing a role. Only when this is established is it pos-sible to ascertain *how* climate change is altering the behaviour of the natural hazard. Unfortunately it is common to attempt to answer the latter (*how*) with-out knowing the former (*if*) or assume that *how* is *if* when relying solely upon the short historical record.

Long-term records of extreme events offer insights into the behaviour of nat-ural hazards beyond that possible from short historical records alone. Already numerous techniques for extracting and deciphering these natural records exist and the data derived highlight that non-randomness or serial correlation is a key characteristic of such events over longer time periods. Recognition of this cyclic-ity and the lack of its incorporation into hazard risk assessments to date suggests that many communities may be unnecessarily exposed to hazard impacts at least in terms of perceived levels of community safety. Hazard risk assessments need to incorporate variable temporal probabilities, determine the uncertainty margins of such estimates and err on the side of caution. Variable temporal probabilities also need to be considered when investigating the role of human-induced cli-mate change on the behaviour of natural hazards. Hopefully, the investigation of long-term records will be a corner stone in the near future in all natural hazard risk assessments. Only then can we hope to understand the true behaviour of the hazard in question.

Appendix

Dating techniques

Radiocarbon dating

Radiocarbon dating relies upon the decay of ^{14}C, relative to the concentration of ^{12}C, within an organism after its death. ^{14}C is produced in the upper atmosphere by the bombardment of ^{14}N by cosmic radiation. The ^{14}C is then transported to the Earth's surface by atmospheric activity such as storms and becomes fixed in the biosphere. It becomes attached to complex organic molecules through photosynthesis in plants and animals ingesting those plants in turn to absorb the ^{14}C along with ^{12}C and ^{13}C. When the organism dies, the ratio of ^{14}C within its carcass begins to gradually decrease. The ^{14}C half-life is 5730 years which means that half of the ^{14}C present at the time of death will decay over this time period.

The ratio of ^{14}C to ^{12}C in the atmosphere has not remained constant over time. This is due to variations in the intensity of the cosmic radiation bombardment of the Earth, and changes in the effectiveness of the Van Allen belts and the upper atmosphere to deflect that bombardment. Comparisons between radiocarbon dates and those from other independent techniques such as dendrochronology and coral growth ring chronologies have allowed the development of calibration tables that outline these variations over time. Radiocarbon dates are calibrated against these tables to convert the result from radiocarbon to calendar years.

Cosmogenic nuclide dating

The method is based upon the same idea as radiocarbon dating of ^{14}C produced in the atmosphere. Cosmogenic nuclides are produced by the interactions of cosmic rays with atoms in a soil or rock surface to produce ^3He, ^{10}Be, ^{14}C, ^{21}Ne, ^{26}Al and ^{36}Cl. ^{36}Cl for example, is produced by spallation reactions of ^{39}K and ^{40}Ca, and by the activation of ^{35}Ca after interactions with cosmic rays. The accumulation rate of these nuclides is proportional to the cosmic ray flux and to the concentration of target nuclides in the surface material. Thus, the concentration of cosmogenic nuclides in sampling data can be used to determine the length of time the sample has spent at or near Earth's surface. The production rates of the nuclides are low, thus they can only be detected using high-resolution mass spectrometers. Accelerator mass spectrometry (AMS) is the most commonly used procedure. AMS uses a particle accelerator combined with a high-resolution detector system that can identify and count single atoms. Cosmogenic nuclide dating can be used to date material ranging from less than 1000 years old to millions of years in age depending on the material that is dated, its age and the analysis facilities that are available. Systematic errors can occur due to the exposure history of the surface being poorly known, the surface having undergone significant erosion during the examined time interval or due to invalid assumptions concerning the isotope production rates. For accurate dating to be possible it is important that samples are taken from unweathered surfaces. It is possible to take samples for dating surface exposure if sufficient rock volume has been removed to expose previously shielded rock surfaces. This method has been successfully applied to the dating of rock fall events, in addition to the dating of glacial deposits, impact craters, basalt flows and other volcanic features and measurements of erosion rates.

Optically stimulated luminescence (OSL) dating

OSL dating can be used to determine the age of aeolian, fluvial, colluvial and littoral sediments. Luminescence is caused by stimulation of electrons trapped in metastable energy levels at crystal defects. These electrons subsequently recombine in luminescence centres. This process leads to the emission of photons when the electrons are returned to shallower trap levels. The electrons can be released when the mineral grain is exposed to light or heat. Thus, luminescence describes emissions of photons over and above black-body radiation by a non-conducting solid to which energy is supplied. The OSL clock is reset in a mineral when it is last exposed to sunlight long enough for all the trapped electrons to be released from the crystal lattice. It has been found that only a few seconds of exposure can reduce the signal to below the detection level. The necessary time for exposure varies with radiation levels. Aeolian sediments have usually been well exposed to sunlight and the mineral grains often have the previously stored luminescence energy released during transport. Fluvial sediments are also often quite well bleached of previous luminescence energy but at times turbid flood flows and those that occur specifically at night can contain mineral grains that have not been well bleached. There are a variety of techniques used in order to test for adequate bleaching and resolving this problem during the dating process. Samples from landslide events can be obtained from the shear surface and at the accumulation lobe. Dating at the shear surface is possible where the energy release of the mass movement was great enough to heat the material at the shear surface to temperatures of 500 °C or more or where frictionites were formed. Dating of material from the accumulation lobe can possibly be achieved in debris flows, mudflows and rock avalanches that contain sandy or silty material if the mineral grains were exposed to sunlight for a sufficient time period to reset the OSL clock.

Uranium-series dating

Uranium-series dating is based on radioactive disequilibrium in the decay chains of ^{238}U and ^{235}U. This disequilibrium is created by geochemical fractionation. The amount of time since the end of the fractionation can be measured by the extent of balance accomplished in the decay chains. Thus, the time span since an event that affected the decay chain of these isotopes can be determined accurately. Thermally ionised mass spectrometry (TIMS) allows direct measurement of the concentration of these isotopes. There are very small analytical errors associated with this method and it can be used to date events ranging from less than 1000 years old to more than 500 000 years. This method can be used to estimate an age range of mass movements by dating carbonates or sinter crusts from above or below the body of a landslide, or by dating carbonates infilling cracks related to such events. Carbonates, from raised coral reefs, during earthquakes are also amenable to the U-series technique.

Argon–Argon (Ar–Ar) dating

This dating method is based on the radioactive decay of ^{40}K to ^{40}Ar. The ^{40}Ar content of tephra ash is determined by the time that has elapsed since the mineral cooled below the temperature at which the system became closed and ^{40}Ar is retained. The method utilises laser fusion combined with noble gas spectrometry and is useful for Quaternary deposits as the ^{40}Ar content in these deposits is usually low and difficult to detect through more conventional methods due to the half-life of 1.2×10^{10}. The method is useful for dating tephra layers that have been deposited following major volcanic eruptions and which act as stratigraphic markers in a sedimentary sequence. Hence, a minimum age for landslides and flood deposits for example can be determined where the tephra overlies the event sediments.

Alpha-recoil-track (ART) dating

The method is deemed to be promising in dating the mineral formation of mica and thus allows for the dating of tephras. As with Ar–Ar dating it can be useful in dating stratigraphic chronologies. The method identifies particle tracks (alpha-recoil-tracks) in the crystal lattice by the recoil of the nucleus during the emission of an α-particle. The α-decay of ^{238}U, ^{235}U and ^{232}Th is used to conduct ART dating.

References

Abell, P. I., Hoelzmann, P. and Pachur, H.-J. (1996). Stable isotope ratios of gastropod shells and carbonate sediments of NW Sudan as palaeoclimatic indicators. *Palaeoecology of Africa*, **26**, 33–52.

Abbott, P. L. (1999). *Natural Disasters*. Boston: McGraw-Hill.

Adams, J. (1981). Earthquake-triggered landslides form lakes in New Zealand. *Earthquake Information Bulletin, United States Geological Survey*, **13**(6), 204–215.

  (1996). Paleoseismology in Canada: a dozen years of progress. *Journal of Geophysical Research*, **101**(B3), 6193–6207.

Agee, R. N. (1980). Present climatic cooling and proposed mechanisms. *Bulletin American Meteorological Society*, **61**, 1356–1367.

Agriculture and Resource Management Council of Australia and New Zealand (ARMCANZ). Standing Committee on Agriculture and Resource Management. (2000). *Floodplain Management in Australia: Best Practice Principles and Guidelines*. Melbourne: CSIRO Publishing.

Alexander, D. (1993). *Natural Disasters*. New York: Chapman and Hall, p. 640.

Allan, R. J., Lindesay, J. and D. Parker. (1996). *El Niño/Southern Oscillation and Climatic Variability*, Collingwood: CSIRO Publishing.

Anderson-Berry, L. (2003). Community vulnerability to tropical cyclones: Cairns, 1996–2000. *Natural Hazards*, **30**, 209–232.

Attrep, M. and Orth, C. J. (1993). Global iridium anomaly, mass extinction, and redox change at the Devonian–Carboniferous boundary. *Geology*, **21**, 1071–1074.

Atwater, B. F. (1987). Evidence for great Holocene earthquakes along the outer coast of Washington State. *Science*, **236**, 942–944.

Atwater, B. F. and Moore, A. L. (1992). Tsunami about 1000 years ago in Puget Sound, Washington. *Science*, **258**, 1614–1617.

Australian Antarctic Division (2004). A large volcanic eruption in 1459. Australian Arctic Division. Accessed online (25/5/04) http://www.ant.div.gov.au/default.asp?casid = 1763

Baines, G. B. K. and McLean, R. F. (1976). Sequential studies of hurricane deposit evolution at Funafuti Atoll. *Marine Geology*, **21**, M1–M8.

Baines, G. B. K., Beveridge, P. K. and Maragos, J. E. (1974). Storms and island building at Funafuti Atoll, Ellice Islands. *Proc. 2nd International Coral Reef Symposia* 2, pp. 485–496.

Baker, V. R. (1973). Paleohydrology and sedimentology of Lake Missoula flooding in eastern Washington. *Geological Society of America Special Paper*, **144**, 1–76.

(1994). Geological understanding and the changing environment. *Transactions of the Gulf Coast Association of Geological Societies*, **44**, 1–8.

(2002). High energy megafloods: planetary settings and sedimentary dynamics. *Special Publication of the International Association of Sedimentologists*, **32**, 3–15.

Baker, V. R. and Pickup, G. (1987). Flood geomorphology of Katherine Gorge, Northern Territory, Australia. *Geological Society of America Bulletin*, **98**, 635–646.

Baker, V. R., Strom, R. G., Gulick, V. C. *et al.* (1991). Ancient oceans, ice sheets and the hydrological cycle on Mars. *Nature*, **352**, 589–594.

Barnes D. J and Taylor R. B. (2001). On the nature and causes of luminescent lines and bands in coral skeletons. *Coral Reefs*, **19**, 221–230.

Beget, J. E., Stihler, S. D. and Stone, D. B. (1994). A 500-year-long record of tephra falls from Redoubt Volcano and other volcanoes in upper Cook Inlet, Alaska. *Journal of Vulcanology and Geothermal Research*, **62**, 55–67.

Bell, F. G. (1998). *Environmental Geology: Principles and Practice*. Oxford: Blackwell Science.

(1999). *Geological Hazards, Their Assessment, Avoidance and Mitigation*. New York: Routledge, p. 216.

Berberian, M. and Yeats, R. S. (2001). Contribution of archaeological data to studies of earthquake history in the Iranian Plateau. *Journal of Structural Geology*, **23**, 563–584.

Biasi, G. P., Weldon II, R. J., Fumal, T. E. and Seitz, G. G. (2002). Paleoseismic event dating and the conditional probability of large earthquakes on the southern San Andreas fault, California. *Bulletin of the Seismological Society of America*, **92**(7), 2761–2781.

Biondi, F., Gershunov, A. and Cayan, D. (2001). North Pacific decadal climate variability since AD 1661. *Journal of Climate*, **14**(1), 5–10.

Blaikie, P., Cannon, T., Davis, I. and Wisner, B. (1994). *At Risk: Natural Hazards, People's Vulnerability, and Disasters*. London: Routledge.

Blecker, S. W., Yonker, C. M., Olson, C. G. and Kelly, E. F. (1997). Palaeopedologic and geomorphic evidence for Holocene climate variation, Shortgrass Steppe, Colorado, USA. *Geoderma*, **76**, 113–130.

Blong, R. J. and Goldsmith, R. C. M. (1993). Activity of the Yakatabari mudslide complex, Porgera, Papua New Guinea. *Engineering Geology*, **35**, 1–17.

Bohor, B. F., Betterton, W. J. and Krogh, T. E. (1993). Impact-shocked zircons: discovery of shock-induced textures reflecting increasing degrees of shock metamorphism. *Earth and Planetary Science Letters*, **119**, 419–424.

Bolt, B. A., Horn, W. L., Macdonald, G. A. and Scott, R. F. (1977). *Geological Hazards*, 2nd edn. New York: Springer-Verlag, p. 330.

Bondevik, S., Mangerud, J., Dawson, S., Dawson, A. and Lohne, O. (2003). Record-breaking height for 8000 year old tsunami in the North Atlantic. *EOS*, **84**, 289–300.

Bondevik, S., Svendsen J. I. and Mangerud, J. (1997). Tsunami sedimentary facies deposited by the Storegga tsunami in shallow marine basins and coastal lakes, western Norway. *Sedimentology*, **44**, 1115–1131.

Bonefile, R. and Chalie, F. (2000) Pollen-inferred precipitation time-series from equatorial mountains, Africa, the last 40k year BP. *Global and Planetary Change*, **26**, 25–50

Boose, E. R., Foster, D. R. and Fluet, M. (1994). Hurricane impacts to tropical and temperate forest landscapes. *Ecological Monographs*, **64**, 369–400.

Boto K. and Isdale P. (1985). Fluorescent bands in massive corals result from terrestrial fulvic acid inputs to the nearshore zone. *Nature*, **315**, 396–397.

Bowler, J. M. (1976). Aridity in Australia: Age, origins and expression on aeolian landforms and sediments. *Earth Science Reviews*, **12**, 279–310.

Bretz, J. H. (1928). The channel scabland of eastern Washington. *Geographical Review*, **18**, 446–477.

Bryant, E. A. (2005). *Natural Hazards*. 2nd edn. Hong Kong: Cambridge University Press, p. 294.

(2001). *Tsunami – the Underrated Hazard*. Melbourne: Cambridge University Press, p. 350.

Bryant, E. A. and Nott, J. F. (2001). Geological indicators of large tsunami in Australia. *Natural Hazards*, **24**, 231–249.

Bryant, E. A. and Young, R. W. (1996). Bedrock-sculpturing by tsunami, south coast New South Wales, Australia. *Journal of Geology*, **100**, 565–582.

Bryant, E. A., Young, R. W. and Price, D. M. (1992). Evidence of tsunami sedimentation on the south-eastern coast of Australia. *Journal of Geology*, **100**, 753–765.

Bryant, E. A., Young, R. W., Price, D. M., Wheeler, D. J. and Pease, M. I. (1997). The impact of tsunami on the coastline of Jervis Bay, south-eastern Australia. *Physical Geography*, **18**, 440–459.

Buckle, P. (1999). Redefining community and vulnerability in the context of emergency management. *Australian Journal of Emergency Management*, **13**(4), 21–26.

Bull, W. B. and Brandon, M. T. (1998). Lichen dating of earthquake-generated regional rockfall events, Southern Alps, New Zealand. *Geological Society of America Bulletin*, **110**(1), 60–84.

Byerly, G. R. and Lowe, D. R. (1994). Spinel from Archean impact spherules. *Geochimica et Cosmochimica Acta*, **58**(16), 3469–3486.

Cameron, K. A. and Pringle, P. T. (1991). Prehistoric buried forests of Mount Hood. *Oregon Geology*, **53**, 34–43.

Cannon, T. (2000). Vulnerability analysis and disasters. In *Floods*, vols 1 and 2, ed. D. J. Parker. London: Routledge, p. 312.

Carling, P. A. (1996). Morphology, sedimentology and palaeohydraulic significance of large gravel dunes. Altai Mountains, Siberia. *Sedimentology*, **43**, 647–664.

Carn, S. A. (2000). The Lamogan volcanic field, East Java, Indonesia: physical vulcanology, historic activity and hazards. *Journal of Volcanology and Geothermal Research*, **95**, 81–108.

Carracedo, J. C., Day, S. J., Guillou, H. and Perez Torrado, F. J. (1999). Giant Quaternary landslides in the evolution of La Palma and El Hierro, Canary Islands. *Journal of Volcanology and Geothermal Research*, **94**, 169–190.

Carrera, P. E. and O'Neill, J. M. (2002). Tree-ring dated landslide movements and their relationship to seismic events in southwestern Montana, USA. *Quaternary Research*, **59**, 25–35.

Chalker, B. E. and Barnes, D. J. (1990). Gamma densitometry for the measurement of coral skeletal density. *Coral Reefs*, **4**, 95–100.

Chapman, D. (1999). *Natural Hazards*, 2nd edn. Singapore: Oxford University Press.

Chappell, J. and Grindrod, J. (1984). Chenier plain formation in northern Australia. In *Coastal Geomorphology in Australia*, ed. B. Thom. Sydney: Academic Press.

Chappell, J. and Polach, H. (1991). Post-glacial sea-level rise from a coral record at Huon Peninsula, Papua New Guinea. *Nature*, **349**, 147–149.

Chappell, J., Chivas, A., Rhodes, E. and Wallensky, E. (1983). Holocene palaeo-environmental changes, central to north Great Barrier Reef inner zone. *BMR Journal of Australian Geology and Geophysics*, **8**, 223–235.

Charles-Dominique, P., Blanc, P., Larpin, D., *et al.* (1998). Forest perturbations and biodiversity during the last ten thousand years in French Guiana. *Acta Oecologica*, **19**(3), 259–302.

Charman, D. J. and Grattan, J. (1999). An assessment of discriminate function analysis in identification and correlation of distal Icelandic tephras in the British Isles. In *Volcanoes in the Quaternary, Geological Society of London Special Publication*, vol. **171**. London: Geological Society of London, pp. 147–160.

Chihara, K., Iwanaga, S., Ito, T., *et al.* (1994). Geohistorical development of the Tochiyama landslide in north-central Japan. *Engineering Geology*, **38**, 205–219.

Chivas A., Chappell, J. and Wallensky, E. (1986). Radiocarbon evidence for the timing and rate of island development, beach rock formation and phosphatization at Lady Elliot Island, Queensland, Australia. *Marine Geology*, **69**, 273–287.

Chivas, A. R., DeDeckker, P. and Shelley, M. G. (1985). Strontium content of ostracods indicates lacustrine palaeosalinity. *Nature*, **316**, 251–253.

Cioni, R., Gurioli, L., Sbrana, A. and Vougioukalakis, G. (2000). Precursory phenomena and destructive events related to the late Bronze Age Minoan (Thera, Greece) and AD 79 (Vesuvius, Italy) Plinian eruptions; inferences from the stratigraphy in the archaeological areas. In *The Archaeology of Geological Catastrophes*, ed. W. G. McGuire, D. R. Griffiths, P. L Hancock and I. S. Stewart, *Geological Society of London Special Publication*, vol. **171**. London: Geological Society of London, pp. 123–141.

Clague, J. J. and Bobrowsky, P. T. (1994). Evidence for a large earthquake and tsunami 100–400 years ago on western Vancouver Island, British Columbia. *Quaternary Research*, **41**, 176–184.

Clague, J. J., Hutchinson, I., Mathews, R. W. and Petterson, R. T. (1999). Evidence for late Holocene tsunamis at Catala Lake, British Columbia. *Journal of Coastal Research*, **15**, 45–60.

Cleaveland, M. K. (2000). A 963-year reconstruction of summer (JJA) streamflow in the White River, Arkansas, USA, from tree rings. *The Holocene*, **10**, 1, 33–41.

Cockell, C. S. (1999). Crises and extinction in the fossil record – a role for ultraviolet radiation? *Palaeobiology*, **25**(2), 212–225.

Coleman, C. (1977). Origin of a stranded bar, Hurricane Eloise, Florida. In *Proceedings 2nd Symposium Coastal Sedimentology*, ed. W. F. Tanner. Tallahassee, FL: Florida State University, pp. 221–228.

Collins, E. S., Scott, D. B. and Gayes, P. T. (1999). Hurricane records on the South Carolina coast: can they be detected in the sediment record? *Quaternary International*, **56**, 15–26.

Connell, J. (1978). Diversity in tropical rainforests and coral reefs. *Science*, **199**, 1302–1310.

Connell, J., Hughes, T. P. and Wallace, C. C. (1997). A 30 year study of coral abundance, recruitment and disturbance at several scales in space and time. *Ecological Monographs*, **67**, 461–488.

Cook, E. R. (2000). Nino 3 index reconstruction. International Tree-Ring Data Bank. Southwestern USA Drought Index Reconstruction. International Tree-Ring Data Bank. IGBP PAGES/World Data Center for Paleoclimatology.

Cook, E. R., D'Arrigo, R. D. and Mann, M. E. (2002). A well-verified, multiproxy reconstruction of the winter North Atlantic Oscillation index since A.D. 1400. *Journal of Climate*, **15**, 1754–1764.

Cook, E. R., Meko, D. M., Stahle, D. W. and Cleaveland, M. K. (1996). Tree-ring reconstructions of past drought across the conterminous United States: tests of a regression method and calibration/verification results. In *Tree Rings, Environment, and Humanity*, ed. J. S., Dean, M. Meko and T. W. Swetnam. Tucson, AZ: Radiocarbon, pp. 559–579.

Costa, J. E. (1983). Paleohydraulic reconstruction of flash flood peaks from boulder deposits in the Colorado Front Range. *Geological Society of America Bulletin*, **94**(8), 986–1004.

Crozier, M. J. (1986). *Landslides: Causes, Consequences and Environment*. London: Croom-Helm, p. 252

Crozier, M. J., Deimal, M. S. and Simon, J. S. (1995). Investigation of earthquake triggering for deep-seated landslides, Taranaki, New Zealand. *Quaternary International*, **25**, 65–73.

Currie, R. G. and Fairbridge, R. (1985). Periodic 18.6 year and cyclic 11 year induced drought and flood in northeastern China and some global implications. *Quaternary Science Reviews*, **23**, 109–134.

Darienzo, M. and Peterson, C. (1990). Episodic tectonic subsidence of late Holocene in salt marshes in northern Oregon central Cascadia margin. *Tectonics*, **9**, 1–22.

Davies, O. K. (1999). Pollen analysis of Tulare Lake, California: Great Basin-like vegetation in Central California during the full-glacial and early Holocene. *Review of Palaeobotany and Palynology*, **107**, 249–257.

Davies, P. J. (1983). Reef growth. In *Perspectives on Coral Reefs*, ed. D. J. Barnes. Australia Institute Marine Science. Manuka: Clouston Publishing, pp. 69–106.

Dawson, A. G. (1994). Geomorphological effects of tsunami run-up and backwash. *Geomorphology*, **10**, 83–94.

Dawson, A. G. and Shi, S. (2000). Tsunami deposits. *Pure and Applied Geophysics*, **157**, 875–897.

Dawson, A. G., Long, D. and Smith, D. E., (1988). The Storegga slides: evidence from eastern Scotland for a possible tsunami. *Marine Geology*, **82**, 271–276.

Dawson, A. G., Shi, S., Dawson, S., Takahashi, T. and Shuto, N. (1996). Coastal sedimentation associated with the June 2nd and 3rd 1994 tsunami in Rajegwesi, Java. *Quaternary Science Reviews*, **15**, 901–912.

Day, S. J., Carracedo, J. C., Guillou, H., *et al.* (2000). Comparison and cross-checking of historical, archaeological and geological evidence for the location and type of historical and sub-historical eruptions of multiple-vent oceanic island volcanoes. In *The Archaeology of Geological Catastrophes*, ed. W. G., McGuire, D. R., Griffiths, P. L. Hancock, and I. S., Stewart, *Geological Society of London Special Publication*, vol. **171**. London: Geological Society of London, pp. 281–306.

Dean, W. (1997). Rates, timing, and cyclicity of Holocene eolian activity in north-central United States: evidence from varved lake sediments. *Geology*, **25**(4), 331–334.

Dean, W. E., Ahlbrandt, T. S., Anderson, R. Y. and Bradbury, J. P. (1996). Regional aridity in North America during the middle Holocene. *Holocene*, **6**(2), 145–155.

De Deckker, P. and Correge, T. (1991). Late Pleistocene record of cyclic eolian activity from tropical Australia suggesting the Younger Dryas is not an unusual climatic event. *Geology*, **19**, 602–605.

Deloule, E., Chaussidon, M., Glass, B. P. and Koeberl, C. (2001). U–Pb isotopic study of relict zircon inclusions recovered from Muong Nong-type tektites. *Geochimica et Cosmochimica Acta*, **65**(11), 1833–1838.

Dominey-Howes, D. (1996). Sedimentary deposits associated with the July 9th, 1956 Aegean Sea tsunami. *Physics and Chemistry of the Earth*, **21**, 51–55.

Dominey-Howes, D. T. M., Papodopolous, G. A. and Dawson, A. G. (2000). Geological and historical investigation of the 1650 A.D. Mt Columbo (Thera Island) eruption and tsunami, Aegean Sea, Greece. *Natural Hazards*, **21**, 83–96.

Donn, W. L. and Balachandrian, N. K. (1969). Coupling a moving air-pressure disturbance and the sea surface. *Tellus*, **21**(5), 701–706.

Donnelly, J. P., Roll, S., Wengren, M., *et al.* (2001a). Sedimentary evidence of intense hurricane strikes from New Jersey. *Geology*, **29**, 615–618.

Donnelly, J. P., Smith-Bryant, S., Butler, J., *et al.* (2001b). 700 year sedimentary record of intense hurricane landfalls in southern New England. *Geological Society of America Bulletin*, **113**, 714–727.

Doumas, C. G. (1990). The prehistoric eruption of Thera and its effect in the Eastern Mediterranean. In *Engineering Geology of Ancient Works, Monuments and Historical Sites*, ed. F. Marinos and S. Koukis. Rotterdam: Balkema, pp. 1983–1985.

Driessen, J. and MacDonald, C. F. (2000). The eruption of Santorini and its effects on Minoan Crete. In *The Archaeology of Geological Catastrophes, Geological Society of London Special Publication*, vol. 171. London: Geological Society of London, pp. 81–93.

Dudley, W. and Lee, M. (1998). *Tsunami*. Hawaii: University of Hawaii Press.

Dury, G. H. (1973). Principles of underfit streams. *US Geological Survey Professional Paper*, **452-A**, A1–A67.

Dyer, B. D., Lyalikova, N. N., Murray, D., *et al.* (1989). Role for microorganisms in the formation of iridium anomalies. *Geology*, **17**, 1036–1039.

Earth Impact Database (2003). Accessed online (17/4/04). http://www.unb.ca/passc/ImpactDatabase/

Ehhalt, D. H. and Östlund, H. G. (1970). Deuterium in hurricane Faith 1966: preliminary results. *Journal of Geophysical Research*, **75**, 2323–2327.

Elliott, G. and Worsley, P. (1999). The sedimentology, stratigraphy and ^{14}C dating of a turf-banked solifluction lobe: evidence for Holocene slope instability at Okstindan, Northern Norway. *Journal of Quaternary Science*, **14**(2), 175–188.

Ely, L. L., Enzel, Y., Baker, V. R. and Cayan, D. R. (1993). A 5000 year record of extreme floods and climate change in the southwestern United States. *Science*, **262**, 103–126.

Emanuel, K. (1987). The dependence of hurricane intensity on climate. *Nature*, **326**, 483–485.

Encrenaz, T., Bibring, J. P. and Blanc, M., (1990). *The Solar System*. Berlin: Springer-Verlag.

Evanc, N. J., Gregorie, D. C., Goodfellow, W. D., *et al.* (1993). Ru/Ir ratios at the Cretaceous–Tertiary boundary: implications for PGE source and fractionation within the ejecta cloud. *Geochimica et Cosmochimica Acta*, **57**, 3149–3158.

Fairbridge, R. (1984). The Nile floods as a global climatic/solar proxy. In *Climatic Changes on a Yearly to Millennial Basis*, ed. N. A. Morner and W. Karle'n. Dordrecht: Reidel, pp. 181–190.

Fairbridge, R. and Fougere, P. F. (1984). Eighty-eight year periodicity in solar terrestrial phenomena confirmed. *Journal of Geophysical Research*, **89**, 413–416.

Fantucci, R. and Sorriso-Valvo, M. (1999). Dendromorphological analysis of a slope near Lago, Calabria (Italy). *Geomorphology*, **30**, 165–174.

FEMA (2002). Guidelines and specifications for flood hazard mapping partners, *Appendix D: guidance for coastal flooding analyses and mapping*. US, Federal Emergency Management Agency. Accessed online [17/6/03] http://www.fema.gov/fhm/en_cfhtr.shtm

Fletcher, C. H., Richmond, B. M., Barnes, G. M. and Schroeder, T. A. (1995). Marine flooding on the coast of Kaua'i during Hurricane Iniki: hindcasting inundation components and delineating washover. *Journal of Coastal Research*, **11**(1), 188–204.

Ford, A. and Rose, W. I. (1995). Volcanic ash in ancient Maya ceramics of the limestone lowlands: implications for prehistoric volcanic activity in the Guatemala highlands. *Journal of Vulcanology and Geothermal Research*, **66**, 149–162.

Forman, S. L., Ogelsby, R. and Webb, R. S. (2001). Temporal and spatial patterns of Holocene dune activity on the Great Plains of North America: megadroughts and climate links. *Global and Planetary Change*, **29**, 1–29.

Fothergill, A. (1996). Gender, risk and disaster. *International Journal of Mass Emergencies and Disasters*, **14**(1), 33–56.

Fothergill, A., Maestas, E. and Darlington, J. (1999). Race, ethnicity and disasters in the United States: a review of the literature. *Disasters*, 23(2), 156–173.

Fournier d'Albe, E. M. (1986). Introduction: reducing vulnerability to nature's violent forces: cooperation between scientist and citizen. In *Violent Forces of Nature*, ed. R. H. Maybury. Maryland: Lomond Publications, pp. 1–6.

French, J. G and Holt, K. W. (1989). Floods. In *The Public Health Consequences of Disasters*, ed. M. B. Greg *et al.*, Chapter 10. Atlanta, GA: United States Department of Health and Human Services, Public Health Services for Disease Control and Prevention.

Fujiwara, O., Matsuda, F., Sakai, T., Irizuki, T. and Fuse, K. (2000). Tsunami deposits in Holocene bay mud in southern Kanto region, pacific coast of central Japan. *Sedimentary Geology*, **135**, 219–230.

Fumal, T. E., Rymer, M. J. and Seitz, G. G. (2002a). Timing of large earthquakes since AD 800 on the Mission Creek Strand of the San Andreas Fault zone at Thousand Palms Oasis near Palm Springs, California. *Bulletin of the Seismological Society of America*, **92**(7), 2841–2860.

Fumal, T. E., Weldon II, R. J., Biasi, G. P., *et al.* (2002b). Evidence for large earthquakes on the San Andreas fault at the Wrightwood, California, paleoseismic site: AD 500 to present. *Bulletin of the Seismological Society of America*, **92**(7), 2726–2760.

Gadbury, C., Todd, L., Jahren, A. H. and Amundson, R. (2000). Spatial and temporal variations in the isotopic composition of bison tooth enamel from the early Holocene Hudson–Meng Bone Bed, Nebraska. *Palaeogeography, Palaeoclimatology, Palaeoecology*, **157**, 79–93.

Gaiser, E., Philippi, T. and Taylor, B. (1998). Distribution of diatoms among intermittent ponds on the Atlantic Coastal Plain: development of a model to predict drought periodicity from surface-sediment assemblages. *Journal of Paleolimnology*, **20**(1), 71–90.

Gamble, J. A., Price, R. C., Smith, I. E. M., McIntosh, W. C. and Dunbar, N. W. (2003). $^{40}Ar/^{39}Ar$ geochronology of magmatic activity, magma flux and hazards at Ruaphehue volcano, Taupo Volcanic Zone, New Zealand. *Journal of Vulcanology and Geothermal Research*, **120**, 271–287.

Giles, T. M., Newnham, R. M., Lowe, D. J. and Munro, A. J. (1999). Impact of tephra fall and environmental change: a 1000 year record from Matakana Island, Bay of Plenty, North Island, New Zealand. *Volcanoes in the Quaternary, Special Publication of the Geological Society of London*, vol. 161. London: Geological Society of London, pp. 11–26.

Gillespie, R., Street-Perrot, F. A. and Switsur, R. (1983). Post-glacial acid episodes in Ethiopia have implications for climate prediction. *Nature*, **306**, 680–683.

Gillieson, D., Ingle Smith, D., Greenaway, M. and Ellaway, M. (1991). Flood history of the limestone ranges in the Kimberley region, Western Australia. *Applied Geographer*, **11**, 105–123.

Glass, B. P. and Wu, J. (1993). Coesite and shocked quartz discovered in the Australasian and North American microtektite layers. *Geology*, **21**, 435–438.

Glikson, A. Y. (1993). Asteroids and early Precambrian crustal evolution. *Earth-Science Reviews*, **35**, 285–319.

Goff, J. and Chagué-Goff, C. (1999). A late Holocene record of environmental changes from coastal wetlands: Abel Tasman National Park, New Zealand. *Quaternary International*, **56**, 39–51.

Goff, J., Chagué-Goff, C. and Nichol, S. (2001). Palaeotsunami deposits: a New Zealand perspective. *Sedimentary Geology*, **143**, 1–6.

Goff, J., Crozier, M., Sutherland, V., Cochran, U. and Shane, P. (1998). Possible tsunami deposit from the 1855 earthquake, North Island, New Zealand. In *Coastal Tectonics*, ed. I. S Stewart and C. Vita-Finzi, *Special Publication of the Geological Society of London*, vol. 146. London: Geological Society of London, pp. 353–374.

Goff, J., McFadgen, B. and Chagué-Goff, C. (2004). Sedimentary differences between the 2002 Easter storm and the 15th Century Okoropunga tsunami, southeastern North Island, New Zealand. *Marine Geology*, **204**, 235–250.

Goff, J., Rouse, H. L., Jones, S. L., *et al.* (2000). Evidence for an earthquake and tsunami about 3100–3400 years ago, and other catastrophic saltwater inundations recorded in a coastal lagoon, New Zealand. *Marine Geology*, **170**, 231–249.

Gonzales, S., Jones, J. M. and Williams, D. L. (1999). Characterisation of tephras using magnetic properties: an example from SE Iceland. In *Volcanoes in the Quaternary, Geological Society of London Special Publication*, vol. 161. London: Geological Society of London, pp. 125–145.

Gonzalez, S., Pastrana, A., Siebe, C. and Duller, G. (2000). Timing of the prehistoric eruption of Xitle Volcano and the abandonment of Cuicuilco Pyearamid, Southern Basin of Mexico. In *The Archaeology of Geological Catastrophes, Geological Society of London Special Publication*, vol. 171. London: 205–224.

Grajales-Nishimura, J. M., Cedillo-Pardo, E., Rosales-Dominguez, C., *et al.* (2000). Chixulub impact: the origin of reservoir and seal facies in the southeastern Mexico oil fields. *Geology*, **28**(4), 307–310.

Granger, K., Jones, T., Leiba, M. and Scott, G. (1999). Community risk in Cairns – a multi-hazard risk assessment. Australian Geological Survey Organization Report.

Grattan, J., Gilbertson, D. and Charman, D. (1999). Modelling the impact of Icelandic volcanic eruptions upon the prehistoric societies and environment of northern and western Britain. In *Volcanoes in the Quaternary, Geological Society of London Special Publication*, vol. 161. London: Geological Society of London, pp. 109–124

Graumlich, L. J., Pisaric, M. F. J., Waggoner, L. A., Littell, J. S. and King, J. C. (2003). Upper Yellowstone river flow and teleconnections with Pacific Basin climate variability during the past three centuries. *Climatic Change*, 59(1–2), 245–262.

Greene, M. T. (1992). *Natural Knowledge in Preclassical Antiquity*. Baltimore, MD: John Hopkins University Press.

Grieve, R. A. F, Langenhorst, F. and Stoffler, D. (1996). Shock metamorphism of quartz in nature and experiment: II. Significance in Geoscience. *Meteoritics and Planetary Science*, 31, 6–35.

Griffiths, D. R. (2000). Uses of volcanic products in antiquity. In *The Archaeology of Geological Catastrophes*, ed. W. G. McGuire, D. R. Griffiths, P. L. Hancock and I. S. Stewart, *Geological Society of London Special Publication*, vol. 171. London: Geological Society of London, pp. 15–23.

Grissino-Mayer, H. D. (1996). A 2129 year annual reconstruction of precipitation for northwestern New Mexico, USA. In *Tree Rings, Environment, and Humanity, Radiocarbon 1996*, ed. J. S. Dean, D. M. Meko and T. W. Swetnam. Tucson, AZ: Department of Geosciences, University of Arizona, pp. 191–204.

Gröcke, D. R., Bocherens, H. and Mariotti, A. (1997). Annual rainfall and nitrogen-isotope correlation in macropod collagen: application as a palaeoprecipitation indicator. *Earth and Planetary Science Letters*, 153, 279–285.

Gucsik, A., Koeberl, C., Brandstatter, F., Reimold, W. U. and Libowitzky, E. (2002). Cathodoluminescence, electron microscopy, and Raman spectroscopy of experimentally shock-metamorphosed zircon. *Earth and Planetary Science Letters*, 202, 495–509.

Hameed, S., Yeh, W. M., Li, M. T., Cess, R. D. and Wang, W. C. (1983). An analysis of periodicities in the 1470–1974 Beijing precipitation record. *Geophysical Research Letters*, 10, 436–439.

Harper, B. (1998). Storm tide threat in Queensland: history, prediction and relative risks. Queensland Department Environment and Heritage, Technical Report 10.

Hassler, S. W. and Simonson, B. M. (2001). The sedimentary record of extraterrestrial impacts in deep-shelf environments: evidence from the early Precambrian. *Journal of Geology*, 109, 1–19.

Hayne, M. and Chappell, J. (2001). Cyclone frequency during the last 5000 years from Curacoa Island, Queensland. *Palaeogeography, Palaeoclimatology, Palaeoecology*. 168, 201–219.

(in press). A record of cyclone frequency from Princess Charlotte Bay, Nth Qld, Australia. *Palaeogeography, Palaeoclimatology, Palaeoecology*.

Heine, K. (2004). Flood reconstructions in the Namib Desert, Namibia and Little Ice Age climate implications: evidence from slackwater deposits and desert soil sequences. *Journal of the Geological Society of India*, 64, 535–547.

Hemphill-Haley, E. (1996). Triatoms as an aid in identifying late Holocene tsunami deposits. *The Holocene*, **6**, 439–448.

Hemphill-Haley, M. A. and Weldon, R. J. (1999). Estimating prehistoric earthquake magnitude from point measurements of surface rupture. *Bulletin of the Seismological Society of America*, **89**(5), 1264–1279.

Hesse, P. (1994). The record of continental dust from Australia in Tasman Sea sediments. *Quaternary Science Reviews*, **13**, 257–272

Hesse, P. and McTainsh, G. H. (1999). Last glacial maximum to early Holocene wind strength in the mid-latitudes of the southern hemisphere from aeolian dust in the Tasman Sea. *Quaternary Research*, **52**, 343–349.

Hewitt, K. (2001). Catastrophic rockslides and the geomorphology of the Hunza and Gilgit river valleys, Karakoram Himalaya. *Erdkunde*, **55**, 72–93.

Hodell, D. A., Curtis, J. H. and Brenner, M. (1995). Possible role of climate in the collapse of the Classic Maya Civilization. *Nature*, **375**, 391–394.

Hoelzmann, P., Kruse, H-J. and Rottinger, F. (2000). Precipitation estimates for the eastern Saharan palaeomonsoon based on a water balance model of the West Nubian Palaeolake Basin. *Planetary and Global Change*, **26**, 105–120.

Holland, G. (1997). The maximum potential intensity of tropical cyclones. *Journal of Atmospheric Science*, **54**, 2519–2541.

Holland, G. and McBride, J. (1997). Tropical Cyclones. In *Windows on Meteorology: Australian Perspective*, ed. E. Webb. Collingwood: CSIRO Publishing, pp. 200–216.

Holmes, J. A., Street-Perrott, F. A., Stokes, S., *et al.* (1999). Holocene landscape evolution of the Manga Grasslands, NE Nigeria: evidence from paleolimnology and dune chronology. *Journal of the Geological Society London*, **156**, 357–368.

Hong, Y. T., Hong. B., Lin, Q. H., *et al.* (2003). Correlation between Indian Ocean summer monsoon and the North Atlantic climate during the Holocene. *Earth and Planetary Science Letters*, **211**, 371–380.

Hong, Y. T., Wang, Z. G., Jiang, H. B., *et al.* (2001). A 6000-year record of changes in drought and precipitation in northeastern China based on a (δ^{13}C) time series from peat cellulose. *Earth and Planetary Science Letters*, **185**, 111–119.

Hubbard, R. N. L. B. and Sampson, C. G. (1993). Rainfall estimates derived from the pollen content of modern hyrax dung: an evaluation. *Suid-Afrikaanse Tydskrif vir Wetenskap*, **89**, 199–204.

Hughes, M. K. and Graumlich, L. J. (1996). Climatic variations and forcing mechanisms of the last 2000 years. In *Multi-millennial Scale Dendroclimatic Studies from the Western United States*, NATO ASI Series, vol. 141. Berlin: Springer-Verlag, pp. 109–124.

Hughes, M. K., Wu, X. D., Shao, X. M. and Garfin, G. M. (1994). A preliminary reconstruction of rainfall in north-central China since A.D. 1600 from tree-ring density and width. *Quaternary Research*, **42**, 88–99.

Hughes, T. P. (1989). Community structure and diversity of coral reefs: the role of history. *Ecology*, **7**, 275–279.

(1999). Off-reef transport of coral fragments at Lizard Island, Australia. *Marine Geology*, **157**, 1–6.

Hughes, T. P. and Connell, J. (1999). Multiple stressors on coral reefs: a long-term perspective. *Limnology and Oceanography*, **44**, 932–940.

Isdale P. J. (1984). Fluorescent bands in massive corals record centuries of coastal rainfall. *Nature*, **310**, 578–579.

Isdale, P. J., Stewart, B. J., Tickle, K. S. and Lough, J. M. (1998). Palaeohydrological variation in a tropical river catchment: a reconstruction using fluorescent bands in corals of the Great Barrier Reef, Australia. *The Holocene*, **8**, 1–8.

Jacoby, G. C., Williams, P. L. and Buckley, B. M. (1992). Tree ring correlation between prehistoric landslides and abrupt tectonic events in Seattle, Washington. *Science*, **258**, 1621–1623.

Jha-Barnouin, O. S. and Schultz, P. H. (1998). Lobateness of impact ejecta deposits from atmospheric interactions. *Journal of Geophysical Research*, **103**(E11), 25 739–25 756.

Jibson, R. W. and Keefer, D. K. (1994). Analysis of the origin of landslides in the New Madrid Seismic Zone. *US Geological Survey Professional Paper*, **1538-D**.

Jones, B. W. (1999). *Discovering the Solar System*. Chichester: John Wiley and Sons.

Jones, J. A. (2002). The physical causes and characteristics of floods. In *Floods*, vol. II, ed. D. J. Parker, London: Routledge.

Kamo, S. L. and Krogh, T. E. (1995). Chicxulub crater source for shocked zircon crystals from the Cretaceous–Tertiary boundary layer, Saskatchewan: evidence from new U–Pb data. *Geology*, **23**(3), 281–284.

Karlin, R. E. and Abella, S. E. B. (1996). Paleoearthquakes in the Puget Sound Region recorded in sediments from Lake Washington, USA. *Science*, **258**, 1617–1620.

Kawata, Y., Borrero, J., Davies, H., *et al.* (1999). Tsunami in Papua New Guinea was as intense as first thought. *EOS Transactions*, AGU, **80**, 101–105.

Keen, T. R. and Slingerland, R. L. (1993). Four storm-event beds and the tropical cyclones that produced them: a numerical hindcast. *Journal of Sedimentary Petrology*, **63**, 218–232.

Keys, C. (1991). Community analysis, some considerations for disaster preparedness and response. *Macedon Digest*, **6**(2), 13–16.

Knighton, D. (1984). *Fluvial Forms and Processes*. London: Edward Arnold, p. 218.

Knox, J. C. (1993). Large increases in flood magnitude in response to modest changes in climate. *Nature*, **361**, 430–432.

(2003). North American palaeofloods and future floods: responses to climatic change. In *Palaeohydrology – Understanding Global Change*, ed. K. J. Gregory and G. Benito. Chichester: John Wiley and Sons, p. 396.

Kochel, R. C. and Baker, V. R. (1988). Paleoflood analysis using slackwater deposits. In *Flood Geomorphology*, ed. V. R. Baker, V. R. Kochel and P. C. Patton. New York: John Wiley and Sons.

Koeberl, C. and Shirey, S. B. (1997). Re–Os isotope systematics as a diagnostic tool for the study of impact craters and distal ejecta. *Palaeogeography, Palaeoclimatology, Palaeoecology*, **132**, 25–46.

Koeberl, C., Reimold, W. U. and Boer, R. H. (1993). Geochemistry and mineralogy of Early Archean spherule beds, Barberton Mountain Land, South Africa:

evidence for origin by impact doubtful. *Earth and Planetary Science Letters*, **199**, 441–452.

Krastel, S., Shmincke. H-U., Jacobs, C. L., *et al.* (2001). Submarine landslides around the Canary Islands. *Journal of Geophysical Research*, **106**(B3), 3977–3997.

Krogh, T. E., Kamo, A. and Bohor, B. F. (1993). Fingerprinting the K/T impact site and determining the time of impact by U–Pb dating of single shocked zircons from distal ejecta. *Earth and Planetary Science Letters*, **119**, 425–429.

Kwiatkowski, D., Phillips, P. C. B., Schmidt, P. and Shin, Y. (1992). Testing the null hypothesis of stationarity against the alternative of a unit root. *Journal of Econometrics*, **54**, 159–178.

Lafferty III, R. H. (1996). Archaeological techniques of dating ancient quakes. *Geotimes*, **41**, 24.

Laird, K. R., Fritz, S. C. and Cumming, B. F. (1998). A diatom based reconstruction of drought intensity, duration and frequency from Moon Lake, North Dakota: a sub-decadal record of the last 2300 years. *Journal of Paleolimnology*, **19**, 161–179.

Laird, K. R, Fritz, S. C., Maasch, K. A. and Cumming, B. F. (1996). Greater drought intensity and frequency before AD 1200 in the northern Great Plains, USA. *Nature*, **384**, 552–554.

Landsea, C. (2000). Climate variability of tropical cyclones: past, present and future. In *Storms 2000*, ed. R. Pielke Jnr. and R. Pielke Sr. New York: Routledge, pp. 220–241.

Lang, A., Moya, J., Corominas, J., Schrott, L. and Dikau, R. (1999). Classic and new dating methods for assessing the temporal occurrence of mass movements. *Geomorphology*, **30**, 33–52.

Lawrence, J. R. and Gedzelman, S. D. (1996). Low stable isotope ratios of tropical cyclone rains. *Geophysical Research Letters*, **23**, 527–530.

Lecointre, J. A., Neall, V. E., Wallace, R. C. and Prebble, W. M. (2002). The 55- to 60 ka Te Whaiau Formation: a catastrophic, avalanche induced, cohesive debris-flow deposit from Pronto-Tongariro Volcano, New Zealand. *Bulletin of Volcanology*, **63**, 509–525.

Leroux, H., Reimold, W. U., Koeberl, C., Horneman, U. and Doukan, J.-C. (1999). Experimental shock deformation in zircon: a transmission electron microscopic study. *Earth and Planetary Science Letters*, **169**, 291–301.

Li, Y., Craven, J., Schweig, E. S. and Obermeier, S. F. (1996). Sand boils induced by the 1993 Mississippi River flood: could they one day be misinterpreted as earthquake-induced liquefaction? *Geology*, **24**(2), 171–174.

Lienkaemper, J. J., Schwartz, D. P., Kelson, K. I., *et al.* (1999). Timing of palaeoearthquakes on the northern Hayward fault – preliminary evidence in El Cerrito, California. United Sates Geological Survey, Open File Report 99–318, p. 35.

Linsley, B. K., Wellington, G. M. and Schrag, D. P. (2000). Decadal sea surface temperature variability in the sub-tropical South Pacific from 1726 to 1997 A.D. *Science*, **290**, 1145–1148.

Liu, K. and Fearn, M. (1993). Lake sediment record of late Holocene hurricane activities from coastal Alabama. *Geology*, **21**, 793–796.

(2000). Reconstruction of prehistoric landfall frequencies of catastrophic hurricanes in northwestern Florida from lake sediment records. *Quaternary Research*, **54**, 238–245.

(2002). Lake sediment evidence of coastal geologic evolution and hurricane history from Western Lake, Florida: reply to Otvos. *Quaternary Research*, **57**, 429–431.

Liu, K., Shen, C. and Louie, K. (2001). A 1000 year history of typhoon landfalls in Guangdong, southern China, reconstructed from Chinese historical documentary records. *Annals of the Association of American Geographers*, **91**, 453–464.

Liu, T., Tan, M. and Qin, X (1997). Climate variations near Beijing over the last 1130 years: evidence from annual layers in a stalagmite from Shihua Cave, China. *Proceedings 19th International Geological Congress*, 2 and 3, pp. 327–332.

Loagicia, H. A., Haston, L. and Michaelsen, J. (1993). Dendrohydrology and long-term hydrological phenomena. *Reviews of Geophysics*, **31**(2), 151–171.

Lockwood, J. P. and Lipman, P. W. (1980.) Recovery of datable charcoal beneath young lavas: lessons from Hawaii. *Bulletin of Volcanology*, **43**, 3.

Losada, M. and Gimenez-Curto, L. (1981). Flow characteristics on rough, permeable slopes under wave action. *Coastal Engineering*, **4**, 187–206.

Lough, J. M. and Barnes, D. J. (1997). Several centuries of variation in skeletal extension, density and calcification in massive Porites colonies from the Great Barrier Reef: a proxy for seawater temperature and a background of variability against which to identify unnatural change. *Journal of Experimental Marine Biology and Ecology*, **211**, 29–67.

(2000). Environmental controls on growth of the massive coral Porites. *Journal of Experimental Marine Biology and Ecology*, **245**, 225–243.

Lough, J. M., Banes, D. J. and McAuister, E. A. (2002). Luminescent lines in corals from the Great Barrier Reef provide spatial and temporal records of reefs affected by land runoff. *Coral Reefs*, **21**, 333–343.

Lowe, D. J. and deLange, W. P. (2000). Volcano–meteorological tsunamis, the *c.* AD 200 Taupo eruption (New Zealand) and the possibility of a global tsunami. *The Holocene*, **10**(3), 401–407.

Lowe, D. R. and Byerly, G. R. (1986). Early Archean silicate spherules of probable impact origin, South Africa and Western Australia. *Geology*, **14**, 83–86.

MacDougal, J. D. (1988). Seawater strontium isotopes, acid rain, and the Cretaceous–Tertiary boundary. *Science*, **239**, 485–487.

Malmquist, D. (1997). Tropical cyclones and climate variability: a research agenda for the next century. Risk Prediction Initiative, Bermuda Biological Station for Research, Bermuda p. 46.

Maragos, J., Baines, G. and Beveridge, P. (1973). Tropical cyclone Bebe creates a new land formation on Funafuti Atoll. *Science*, **181**, 1161–1164.

Matsuda, J. (2000). Seismic deformation of the post 2300 yr BP muddy sediments in Kawachi lowland plain, Osaka, Japan. *Sedimentary Geology*, **135**, 99–116.

McCloskey, J., Nalbant, S. S. and Steacy, S. (2005). Earthquake risk from co–seismic stress. *Nature*, **434**, p. 291.

McGuire, J. J., Boettcher, M. S. and Jordan, T. H. (2005). Foreshock sequences and short-term earthquake predictability on East Pacific Rise transform faults. *Nature*, **434**, 457–461.

McInnes, K. L., Walsh, K. J. E., Hubbert, G. D. and Beer, T. (2003). Impact of sea-level rise and storm surges on a coastal community. *Natural Hazards*, **30**, 187–207.

McInnes, K., Walsh, K. and Pittock, B. (1999). Impact of sea-level rise and storm surges on coastal resorts. A report for CSIRO Tourism Research, CSIRO Atmospheric Research, Melbourne, Australia.

McKee, E. D. (1959). Storm sediments on a Pacific atoll. *Journal of Sedimentary Petrology*, **29**, 354–364.

McLean, R. F. (1993). A two thousand year history of low latitude tropical storms, preliminary results from Funafuti Atoll, Tuvalu. *Proceedings 7th International Coral Reef Symposium*.

McSaveney, M. J., Goff, J. R, Darby, D. A., *et al.* (2000). The July 1998 Papua New Guinea tsunami; evidence and initial interpretation. *Marine Geology*, **170**, 81–92.

Meghraoui, M., Gomez, F, Sbeinati, R., *et al.* (2003). Evidence for 830 years of seismic quiescence from palaeoseismology, archaeoseismology and historical seismicity along the Dead Sea fault in Syearia. *Earth and Planetary Science Letters*, **210**, 35–52.

Meko, D. M., Therrell, M. D., Baisan, C. H. and Hughes, M. K. (2001). Sacramento river flow reconstructed to A.D. 869 from tree rings. *Journal of the American Water Resources Association*, **37**, 4–9.

Ming, T., Dongsheng, L., Xiaouguang, Q., *et al.* (1997). Preliminary study on the data from microbanding and stable isotopes of stalgamites of Bejing Shihua cave. *Carsologica Sinica*, **16**, 1.

Minoura, K. and Nakaya, T. (1991). Traces of tsunami preserved in intertidal lacustrine and marsh deposits, some examples from northeast Japan. *Journal of Geology*, **99**, 265–287.

(1994). Discovery of an ancient tsunami deposit in coastal sequences of southwest Japan: verification of large historic tsunami. *The Island Arc*, **3**, 66–72.

Minoura, K., Imamura, F., Kuran, U., *et al.* (2000). Discovery of Minoan tsunami deposits. *Geology*, **28**, 59–62.

Minoura, K., Imamura, F., Takahashi, T. and Shuto, M. (1997). Sequence of sedimentation processes caused by the 1992 Flores tsunami: evidence from Babi Island. *Geology*, **25**, 523–526.

Montanari, A., Asaro, F., Michel, H. V. and Kennet, J. P. (1993). Iridium anomalies of Late Eocene age at Massignano (Italy), and ODP Site 689B (Maud Rise, Antarctica). *Palaios*, **8**, 420–437.

Mullins, H. T. (1998). Holocene lake level and climate change inferred from marl stratigraphy of the Cayuga Lake Basin, New York. *Journal of Sedimentary Research*, **68**(4), 569–578.

Murck, B. W., Skinner, B. J. and Porter, S. C. (1996). *Environmental Geology*. New York: John Wiley and Sons, p. 535.

Murnane, R. J., Barton, C., Collins, E., *et al.* (2000). Model estimates hurricane wind speed probabilities. *EOS*, **81**, 433–438.

Nanayama, F., Satake, K., Furukawa, R., *et al.* (2003). Unusually large earthquakes inferred from tsunami deposits along the Kuril Trench. *Nature*, **424**, 660–662.

Nanayama, F., Shigeno, K., Satake, K., *et al.* (2000). Sedimentary differences between 1993 Hokkaido-nansei-oki tsunami and 1959 Miyakojima typhoon at Tasai, southwestern Hokkaido, northern Japan. *Sedimentary Geology*, **135**, 255–264.

Nanson, G. C., Chen, X. Y. and Price, D. M. (1992). Lateral migration, thermoluminescence chronology and colour variation of longitudinal dunes near Birdsville in the Simpson Desert, central Australia. *Earth Surface Processes and Landforms*, **17**, 807–819.

NASA (2004). Impact Data Base. Accessed online [17/04/04] http://impact.arc.nasa.gov/index.html

Natawidjaja, D. M. (2002). Neotectonics of the Sumatran Fault and paleogeodesy of the Sumatran subduction zone. PhD thesis (unpublished), California Institute of Technology Pasadena, California.

Neilsen, P. and Hanslow, D. J. (1991). Wave run-up distributions on natural beaches. *Journal Coastal Research*, **7**, 1139–1152.

Nelson, A. R. and Manley, W. F. (1992). Holocene coseismic and aseismic uplift of Isla Mocha, south central Chile. *Quaternary International*, **15–16**, 61–76.

Nichols, N. (1984). The Southern Oscillation, sea surface temperature and interannual fluctuations in Australian tropical cyclone activity. *International Journal of Climatology*, **4**, 661–670.

Nicoll, R. S. and Playford, P. E. (1993). Upper Devonian iridium anomalies, condont zonation and the Frasnian–Famennian boundary in the Canning Basin, Western Australia. *Palaeogeography, Palaeoclimatology, Palaeoecology*, **104**, 105–113.

Nigram, R., Khare, N. and Nair, R. R. (1995). Foraminiferal evidence for 77-year cycles of droughts in India and its possible modulation by the Gleissberg solar cycle. *Journal of Coastal Research*, **11**(4), 1099–1107.

Nikonov, A. A. (1995). The stratigraphic method in the study of large past earthquakes. *Quaternary International*, **25**, 47–55.

Noormets, R., Felton, E. A. and Crook, K. (2002). Sedimentology of rocky shorelines: 2, Shoreline megaclasts on the north shore of Oahu, Hawaii – origins and history. *Sedimentary Geology*, **150**, 31–45.

Norris, R. D., Huber, B. T. and Self-Trail, J. (1999). Synchroneity of the K-T oceanic mass extinction and meteorite impact: Blake Nose, western North Atlantic. *Geology*, **27**(5), 419–422.

Nott, J. F. (1997). Extremely high-energy wave deposits inside the Great Barrier Reef, Australia; determining the cause – tsunami or tropical cyclone. *Marine Geology*, **141**, 193–207.

(2003a). Intensity of prehistoric tropical cyclones. *Journal of Geophysical Research*, **108**(D7), 4212–4223.

(2003b). The importance of prehistoric data and the variability of hazard regimes in natural hazard risk assessment. *Natural Hazards*, **30**, 43–58.

(2003c). Tsunami or storm waves? – Determining the origin of a spectacular field of wave emplaced boulders using numerical storm surge and wave models and hydrodynamic transport equations. *Journal of Coastal Research*, **19**, 348–356.

(2003d). Waves, coastal boulders and the importance of the pre-transport setting. *Earth and Planetary Science Letters*, **210**, 269–276.

(2004). The tsunami hypothesis – comparisons of the field evidence against the effects, on the Western Australian coast, of some of the most powerful storms on Earth. *Marine Geology*, **208**, 1–12.

Nott, J. F. and Bryant, E. A. (2003). Extreme marine inundations (tsunamis?) of coastal Western Australia. *Journal of Geology*, **111**, 691–706.

Nott, J. F. and Hayne, M. (2001). High frequency of 'super-cyclones' along the Great Barrier Reef over the past 5000 years. *Nature*, **413**, 508–512.

Nott, J. and Hubbert, G. (2005). Comparisons between topographically surveyed debris lines and modelled inundation levels from severe tropical cyclones Voice and Chris, and their geomorphic impact on the sand coast. *Australian Meteorological Magazine*, **54**, 187–196.

Nott, J. F. and Price, D. (1994). Plunge-pools and paleoprecipitation. *Geology*, **22**, 1047–1050.

(1999). Waterfalls, floods and climate change: evidence from tropical Australia. *Earth and Planetary Science Letters*, **171**, 267–276.

Nott, J. F., Bryant, E. A. and Price, D. M. (1999). Early Holocene aridity in tropical northern Australia. *The Holocene*, **9**, 231–236.

Otvos, E. G. (2002). Discussion of 'Prehistoric landfall frequencies of catastrophic hurricanes' by Liu and Fearn, 2000. *Quaternary Research*, **57**, 425–428.

Page, K., Nanson, G. and Price, D. (1996). Chronology of Murrumbidgee River palaeochannels on the Riverine Plain, southeastern Australia. *Journal of Quaternary Science*, **11**(4), 311–326.

Pesonen, L. J. (1996). The impact cratering record of Fennoscandia. *Earth, Moon and Planets*, **72**, 377–393.

Pielke, R. A. and Lardsea, C. W. (1998). Normalized hurricane damage in the United States: 1925–95. *Weather and Forecasting*, **13**, 621–631.

Pierrard, O., Robin, E., Rocchia, R. and Montanari, A. (1998). Extraterrestrial Ni-rich spinel in upper Eocene sediments from Massignano, Italy. *Geology*, **26**(4), 307–310.

Pirazolli, P. A., Laborel, J. and Stiros, S. C. (1996). Earthquake clustering in the eastern Mediterranean during historical times. *Journal of Geophysical Research*, **101**(B3), 6083–6087.

Plunket, P. and Urunuela, G. (2000). The archaeology of a Plinian eruption of the Popocatépetl volcano. In *The Archaeology of Geological Catastrophes*, ed. W. G. McGuire, D. R. Griffiths, P. L. Hancock and I. S. Stewart, *Geological Society of London Special Publication*, vol. 171. London: Geological Society of London, pp. 195–203.

Proctor, C. J., Baker, A., Barnes W. L. and Gilmour, M. A. (2000). A thousand year speleothem proxy record of North Atlantic climate from Scotland. *Climate Dynamics*, **16**(10–11), 815–820.

Rabinovich, A. B. and Monserrat, S. (1996). Meteorological tsunamis near the Balearic and Kuril Islands: descriptive and statistical analysis. *Natural Hazards*, **13**, 55–90.

Rasser, M. W. and Riegl, B. (2002). Holocene reef rubble and its binding agents. *Coral Reefs*, **21**, 57–72.

Reimold, W. U., von Brunn, V. and Koeberl, C. (1997). Are diamictites impact ejecta? Supporting evidence from South African Dwyka Group diamictite. *Journal of Geology*, **105**(5), 517–531.

Retallack, G. J., Seyedolali, A., Krull, E. S., *et al.* (1998). Search for evidence of impact at the Permian–Triassic boundary in Antarctica and Australia. *Geology*, **26**(11), 979–982.

Reynolds, L., Rosenbaum, G., Van Metre, C., *et al.* (1999). Greigite (Fe_3S_4) as an indicator of drought; the 1912–1994 sediment magnetic record from White Rock Lake, Dallas, Texas, USA. *Journal of Paleolimnology*, **21**, 193–206.

Rhodes, E. G., Polach, H. A., Thom, B. G. and Wilson, S. R. (1980). Age structure of Holocene coastal sediments, Gulf of Carpentaria, Australia. *Radiocarbon*, **22**, 718–727.

Robin, E., Bonté, P., Froget, L., Jéhanno, C. and Rocchia, R. (1992). Formation of spinels in cosmic objects during atmospheric entry: a clue to the Cretaceous–Tertiary boundary event. *Earth and Planetary Science Letters*, **108**, 181–191.

Robinson, A. (1993). *Earth Shock – Climate, Complexity and the Forces of Nature*. London: Thames and Hudson.

Rondot, J. (1994). Recognition of eroded astroblemes. *Earth-Science Reviews*, **35**, 331–365.

Rudoy, A. N. and Baker, V. R. (1993). Sedimentary effects of cataclysmic late Pleistocene glacial outburst flooding, Altay Mountains, Russia. *Sedimentary Geology*, **85**, 53–62.

Satake, K., Shimazaki, K., Tsuji, Y. and Ueda, K. (1996). Time and size of a giant earthquake in Cascadia inferred from Japanese tsunami records of January 1700. *Nature*, **379**, 246–249.

Sato, H. and Taniguchi, H. (1997). Relationship between crater size and ejecta volume of recent magmatic and phreato-magmatic eruptions: implications for energy partitioning. *Geophysical Research Letters*, **24**(13), 205–208.

Sato, H., Shimamoto, T., Tstsumi, A. and Kawanoto, E. (1995). Onshore tsunami deposits caused by the 1993 southwest Hokkaido and 1983 Japan Sea earthquakes. *Pure and Applied Geophysics*, **144**(3–4), 693–717.

Saucier, R. T. (1991). Geoarchaeological evidence of strong prehistoric earthquakes in the New Madrid (Missouri) seismic zone. *Geology*, **19**, 296–298.

Sawlowicz, Z. (1993). Iridium and other platinum-group elements as geochemical markers in sedimentary environments. *Palaeogeography, Palaeoclimatology, Palaeoecology*, **104**, 253–270.

Scoffin, T. (1993). The geological effects of hurricanes on coral reefs and the interpretation of storm deposits. *Coral Reefs*, **12**, 203–221.

Sheffers, A. (2002). Paleotsunamis in the Caribbean – field evidences and datings from Aruba, Curaçao and Bonaire. *Essener Geographische Arbeiten*, **33**.

Sheppard, P. R. and Jacoby, G. C. (1989). Application of tree-ring analysis to paleoseismology: two case studies. *Geology*, **17**, 226–229.

Shi, S., Dawson, S. and Smith, D. E. (1995). Coastal sedimentation associated with the December 12th 1992 tsunami in Flores Indonesia. *Pure and Applied Geophysics*, **144**, 525–536.

Siebert, L., Beget, J. E. and Glicken, H. (1995). The 1883 and late-prehistoric eruptions of Augustine volcano, Alaska. *Journal of Vulcanology and Geothermal Research*, **66**, 367–395.

Simonson, B. M. and Harnik, P. (2000). Have distal impact ejecta changed over time? *Geology*, **28**(11), 975–978.

Sims, J. D. and Garvin, C. D. (1996). Recurrent liquefaction induced by the 1989 Loma Prieta earthquake and the 1990 and 1991 aftershocks: implications for paleoseismicity studies. *Bulletin of the Seismological Society of America*, **85**(1), 51–65.

Smit, J. (1999). The global stratigraphy of the Cretaceous–Tertiary boundary impact ejecta. *Annual Review of Planetary Sciences*, **27**, 75–113.

Smit, J. and Kyte, F. T. (1985). Cretaceous–Tertiary extinctions: alternative models. *Science*, **230**, 1292–1293.

Smith. A. J., Donovan, J. J., Ito, E. and Engstrom, D. R. (1997). Ground-water processes controlling a prairie lake's response to middle Holocene drought. *Geology*, **25**(5), 391–394.

Smith, K. (2001). *Environmental Hazards, Assessing Risk and Reducing Disaster*. London: Routledge.

Smith, K. and Ward, R. (1998). *Floods: Physical Processes and Human Impacts*. New York: John Wiley and Sons.

Stahle, D. W., *et al.* (2000). Tree-ring data document 16th century megadrought over North America. *EOS*, **81**(12), 121.

Stahle, D. W., Cleaveland, M. K., Blanton, D. B., Therrell, M. D. and Gay, D. A. (1998). The lost colony and Jamestown droughts. *Science*, **280**, 564–567.

Stine, S. (1994). Extreme and persistent drought in California and Patagonia during medieval time. *Nature*, **369**, 546–549.

Stiros, S. C. (1996). Identification of earthquakes from archaeological data: methodology, criteria and limitations. Fitch Laboratory Occasional Paper 7, British School at Athens, Oxford.

Stiros, S. C., Laborel, J. Laborel-Deguen, F., *et al.* (2000). Seismic coastal uplift in a region of subsidence: Holocene raised shorelines of Samos Island, Greece. *Marine Geology*, **170**, 41–58.

Street-Perrot, F. A. and Harrison, S. P. (1984). Temporal variations in lake levels since 30 000 yr BP – an index of the global hydrological cycle. *Geophysical Monographs*, **29**, 118–129.

Stuiver, M. (1980). Solar variability and climatic change during the current millennium. *Nature*, **286**, 868–871.

Stuiver, M. and Quay, P. D. (1980). Changes in atmospheric ^{14}C attributed to a variable Sun. *Science*, **207**, 11–19.

Sturkell, E. F. F. (1998). Resurge morphology of the marine Lockne impact crater, Jämtland, central Sweden. *Geological Magazine*, **135**(1), 121–127.

Sussmilch, C. A. (1912). Note on some recent marine erosion at Bondi. *Proceedings of the Royal Society of New South Wales*, **46**, 155–158.

Tan, M., Liu, J. Hou, X., *et al.* (2003). Cyclic rapid warming on centennial-scale revealed by a 2650-year stalagmite record of warm season temperature. *Geophysical Research Letters*, **30**(12), 1617.

Tanner, W. F. (1995). Origin of beach ridges and swales. *Marine Geology*, **129**, 149–161.

Taylor, M. and Stone, G. W. (1996). Beach ridges: a review. *Journal of Coastal Research*, **12**, 612–621.

Taylor, S. R. (2001). *Solar System Evolution: a New Perspective*. Cambridge: Cambridge University Press.

Thomas, M. F. (2004). Landscape sensitivity to rapid environmental change – a Quaternary perspective with examples from tropical areas. *Catena*, **55**, 107–124.

Tinner, W. and Lotter, A. F. (2001). Central European vegetation response to abrupt climate change at 8.2 ka. *Geology*, **29**(6), 551–554.

Torrence, R., Pavlides, C., Jackson, P. and Webb, J. (2000). Volcanic disasters and cultural discontinuities in Holocene time, in West New Britain, Papua New Guinea. *The Archaeology of Geological Catastrophes, Geological Society of London Special Publication*, vol. 171. London: Geological Society of London, pp. 225–244.

Tuttle, M. P. and Schweig, E. S. (1995). Archaeological and pedological evidence for large prehistoric earthquakes in the New Madrid seismic zone, central United States. *Geology*, **23**(3), 253–256.

(1996). Recognising and dating prehistoric liquefaction features: lessons learned in the New Madrid seismic zone, central United States. *Journal of Geophysical Research*, **101**(B3), 6171–6178.

USGS (1998). Preliminary analysis of sedimentary deposits from the 1998 Papua New Guinea (PNG) tsunami. Accessed online [14/06/05] http://walrus.ws.usgs.gov/tsunami/itst.html

Van't Veer, R., Islebe, G. A. and Hooghiemstra, H. (2000). Climatic change during the Younger Dryas chron in northern South America: a test of the evidence. *Quaternary Science Reviews*, **19**, 1821–1835.

Vasilyev, N. V. (1998). The Tunguska Meteorite problem today. In *Planetary and Space Science*, Special Tunguska Issue, ed. M. Di Hartino, P. Farinella and G. Longo, **46**(2–3), 129–150.

Verschuren, D., Laird, K. R. and Cumming, B. F. (2000). Rainfall and drought in equatorial east Africa during the past 1,100 years. *Nature*, **403**, 410–414.

Wallace, M. W., Keays, R. R. and Gostin, V. A. (1991). Stromatolitic iron oxides: evidence that sea-level changes can cause sedimentary iridium anomalies. *Geology*, **19**, 551–554.

Wasson, R. J. (1986). Geomorphology and Quaternary history of the Australian continental dunefield. *Geographical Review of Japan*, **59**, 55–67.

Wei, W. (1995). How many impact-generated microspherule layers in the Upper Eocene? *Palaeogeography, Palaeoclimatology, Palaeoecology*, **114**, 101–110.

Wells, A., Duncan, R. P. and Stewart, G. H. (2001). Forest dynamics in Westland, New Zealand: the importance of large, infrequent earthquake-induced disturbance. *Journal of Ecology*, **89**, 1006–1018.

Wetherill, G. W. and Shoemaker, E. M. (1982). Collision of astronomically observed bodies with the Earth. In *Geological Implications of Impacts of Large Asteroids and Comets on the Earth*, ed. L. Silver and P. Schultz. Boulder, CO: The Geological Society of America.

White, J. W., Lawrence, J. R. and Broecker, W. S. (1994). Modeling and interpreting D/H ratios in tree rings: a test case of white pine in the northeastern United States. *Geochimica et Cosmochimica Acta*, **58**, 851–862.

Whitehead, J., Papanastassiou, D. A., Spray, J. G., Grieve, R. A. F. and Wasswerburg, G. J. (2000). Late Eocene impact ejecta: geochemical and isotopic connections with Popigai impact structure. *Earth and Planetary Science Letters*, **181**, 473–487.

Wilf, P., Wing, S. L., Greenwood, D. R. and Greenwood, C. L. (1998). Using fossil leaves as palaeoprecipitation indicators: an Eocene example. *Geology*, **26**(3), 203–206.

Williams (1983). Paleohydrological methods and some examples from Swedish fluvial environments 1–cobble and boulder deposits. *Geografiska Annaler*, **65**A, 227–243.

Williams, D. M and. Hall, A. M. (2004). Cliff-top megaclast deposits of Ireland, a record of extreme waves in the North Atlantic – storms or tsunamis? *Marine Geology*, **206**, 101–117.

Williams, G. P. and Costa, J. E. (1988). Geomorphic measurements after a flood. In *Flood Geomorphology*, ed. V. Baker, R. Kochel and P. Patton. New York: John Wiley and Sons, pp. 65–77.

Winter, A., Tadamichi, O., Ishioroshi, H., Watanabe, T. and Christy, J. R. (2000). A two-to-three degree cooling of Caribbean Sea surface temperatures during the Little Ice Age. *Geophysical Research Letters*, **27**(20), 3365–3368.

Wohl, E. (1992a). Bedrock benches and boulder bars: floods in the Burdekin Gorge of Australia. *Geological Society of America Bulletin*, **104**, 770–778.

(1992b). Gradient irregularity in the Herbert Gorge of northeastern Australia. *Earth Surface Processes and Landforms*, **17**, 69–84.

(1994). Sedimentary records of late Holocene floods along the Fitzroy and Margaret Rivers, Western Australia. *Australian Journal of Earth Science*, **41**, 273–280.

(1995). Data for palaeohydrology. In *Global Continental Palaeohydrology*, ed. K. Gregory, L. Starkel and V. Baker. Chichester: John Wiley and Sons, pp. 23–59.

Woodhouse, C. A. and Overpeck, J. T. (1998). 2000 years of drought variability in the central United States. *Bulletin of the American Meteorological Society*, **79**(12), 2693–2714.

Wright, C. and Mella, A. (1963). Modifications to the soil pattern of south-central Chile resulting from seismic and associated phenomena during the period May to August 1960. *Bulletin of the Seismological Society of America*, **53**, 1367–1402.

Young, R. W. (1985). Waterfalls: form and process. *Zeitschrift für Geomorphologie*, **29**, 115–123.

Yeh, H., Imumura, F., Synolakis, C., *et al.* (1993). The Flores tsunami. *EOS, Transactions AGU*, **74**, 369–373.

Young, R. W., Bryant, E. A. and Price, D. M. (1996). Catastrophic wave (tsunami?) transport of boulders in southern New South Wales, Australia. *Zeitschrift fur Geomorphologie, N.F.*, **40**, 191–207.

Zachariasen, J., Sieh, K., Taylor, F. W., Edwards, R. L. and Hantoro, W. S. (1999). Submergence and uplift associated with the giant 1833 Sumatran subduction earthquake: evidence from coral microatolls. *Journal of Geophysical Research*, **104**, 895–919.

Zachariasen, J., Sieh, K., Taylor, F. W. and Hantoro, W. S. (2000). Modern vertical deformation above the Sumatran subduction zone: paleogeodetic insights from coral microatolls. *Bulletin of the Seismological Society of America*, **90**(4), 897–913.

Zebrowski, E. (1997). *Perils of a Restless Planet: Scientific Perspectives on Natural Disasters.* New York: Cambridge University Press.

Ziman, J. (1978). *Reliable Knowledge: an Exploration for the Grounds for Belief in Science.* Cambridge: Cambridge University Press, p. 197.

Index